EARTH LIGHTS REVELATION

EARTH LIGHTS REVELATION

UFOs and Mystery Lightform Phenomena: The Earth's Secret Energy Force

Paul Devereux
with contributing researchers
David Clarke, Andy Roberts
and Paul McCartney

BLANDFORD

Blandford Press
An imprint of Cassell
Villiers House, 41/47 Strand, London WC2N 5JE

First published 1989

This paperback edition first published 1990

Distributed in the United States by
Sterling Publishing Co, Inc,
387 Park Avenue South, New York, NY 10016-8810

Distributed in Australia by
Capricorn Link (Australia) Pty Ltd
PO Box 665, Lane Cove, NSW 2066

British Library Cataloguing in Publication Data
Devereux, Paul, *1945-*
Earth lights revelation: UFO's and mystery lightform phenomena: the Earth's secret energy force.
1. Unidentified flying objects
I. Title
001.9,42

ISBN 0-7137-2209-6

Typeset by Chapterhouse, 10 on 11pt Baskerville,
Formby, L37 3PX, Great Britain

Printed in Great Britain by
Biddles Ltd, Guildford and King's Lynn

CONTENTS

ACKNOWLEDGEMENTS

It is always a risky business listing those persons to whom one owes gratitude for help in producing a book. It is all too easy to miss out someone's name unintentionally. If that has happened here, I beg forgiveness. Any oversight on my part indicates my incompetence, not any lack of gratitude.

For valuable items of information I thank, in no special order, Chris Garner, Phil Rickman, Inga Deland, Bill Becker, Chris Hall, Susan Wood, Bill Rudersdorf, Phyllis Atwater, Andy Collins, Margaret Fry, Jonathan Mullard, Keith Stevenson, Eileen Goodchild, Chris Castle, Paul Bennett, J. Havelock Fidler, George Sandwith, Jeremy Harte, Dan Mattsson, Trish Pfeiffer, Jean Sheridan, John Lobell and various personnel in the British Geological Survey.

For making available priceless and undoubtedly authentic pictures of light phenomena, and in many cases adding important information as well, I thank: Dale Kaczmarek, Jim Crocker, Tony Dodd, Greg Long, Bill Kingsley, Leif Havik and other members of the noble Project Hessdalen, John Merron, David Kubrin, Harley Rutledge. Pictures used from other sources have been identified in the captions, and we all owe a debt of gratitude to those photographers as well. All pictures remain strictly the copyright of the photographers.

My son, Solomon, has saved me a great deal of time in sorting out various computer matters that are beyond my ken, for which I am most appreciative.

John Derr of the US Geological Survey, Michael Persinger of Laurentian University and Brian Brady of the US Bureau of Mines deserve special thanks for their exceptional work in the study of geologically-related light phenomena. They show what science should be all about. I particularly wish to thank Michael Persinger for sending me the full, awesome, set of his research papers, and both he and John Derr for always being prepared to respond to correspondence.

It has been a pleasure working with the contributing researchers to this work. The Pennine material of David Clarke and Andy Roberts is of major importance, and they have also contributed other items of information to this book. (Appreciation must also be given to those – named in Chapter 4 – who have supported Project Pennine. I must single out Philip Mantle for being so helpful with photographic sources.) Paul McCartney and I go back many years, and although personal circumstances prevented his association with this work to the degree he wished, I still want to acknowledge the contributions he has managed to make.

Thanks are due to my editor, Stuart Booth, for being what a good commissioning editor should be: open to, and supportive of, new material, and trusting an author to get on with the job unhindered.

Finally, I owe my wife, Charla, a great debt of gratitude for the way she has been a tower of strength and support in the execution of this project.

P. D.

INTRODUCTION

The subject of this book is an extraordinary type of energy that regularly appears on our planet. This energy is in some way associated with geology, and usually manifests in the form of curious light phenomena – 'earth lights'.

This is not a book 'about UFOs', but earth lights have, alas, become inextricably bound up with the contentious area of ufology, hindering both their perception and unbiased investigation. Thus, we have to deal with aspects of the UFO problem, but the following chapters go on to explore the idea that our own planet produces a range of unknown kinds of lightforms that have been observed down the ages and interpreted in many different ways while never being seen for what they are – expressions either of an energy base of unknown nature or the result of unfamiliar concatenations of known forces. It will be argued that these light phenomena have the potential to affect our understanding of geophysics, the origins of life on Earth, and even aspects of consciousness and brain function. We are looking at a primordial force that possesses the possibility of opening up to us remarkable new energy applications for the twenty-first century.

As a consequence of a decade of unintentionally related research, my first book on this subject, *Earth Lights* (Turnstone Press), was published in 1982, with the collaboration of trained geochemist Paul McCartney. It was the first hardback, full-size book wholly devoted to the topic of the earth light phenomenon to appear in Britain, and, as far as I am aware, only the second such book to be published in the world (during research for *Earth Lights* it was found that *Space-Time Transients and Unusual Events* by Michael Persinger and Gyslaine Lafrenière had been published by Nelson-Hall in 1977). It pulled the material on earth lights together in a way that had not been available previously, as well as providing new examples of regional studies.

In the years since the publication of *Earth Lights* there have been research developments in the subject, some existing material has had further work carried out on it, new evidence has come to light, and thinking around the topic has crystallised to some extent. Important new field or laboratory research has been carried out in, notably, Canada, the USA, Scandinavia and Britain. A most important regional study in Britain, Project Pennine, was instigated as a consequence of the debate within ufology occasioned by the publication of *Earth Lights*. I am pleased that this present book is able to include the first account of that research effort through contributors David Clarke and Andy Roberts, founders of Project Pennine.

This whole range of new or developed material has either not been published previously, or has seen the light of day only fragmentarily in lectures, articles in specialised journals, or as contributory material in disparate publications. This book gathers all the relevant information together, to provide the most comprehensive and up-to-date single source work on earth lights and their possible implications, and to help consolidate the earth lights theory while making it accessible to a wider readership – general readers interested in challenging phenomena at the edges of our current cultural and scientific awareness, as well as those who may have specialised interests in aspects of the material. I have written this work as a non-scientist primarily for non-scientists, but the totality of the material presented should be thoughtfully noted by any scientist worthy of the name.

The first part of the book deals with the emergence into our awareness of curious light forms occurring naturally, if largely inexplicably, on Earth, and places earth lights in that context. We learn how difficult it has been for these phenomena even to be seen for what they are. Part 2 provides what I believe to be a unique geography of earth light phenomena. Much of this material will be unfamiliar to most readers, and some of it has never previously been published. Only by carefully going through these accounts will the reader be able to appreciate fully the skein of characteristics that identify earth lights. Some of the most remarkable accounts of light phenomena ever recorded are presented here. In the third and final section, we attempt to obtain a clearer picture of the energy phenomenon we are confronting, give an account of work being done on the nature of the lights, and finally sketch in some of the lights' possible implications for human understanding. I must make it clear that the ideas put forward in this final, more speculative part of the book do not necessarily reflect those of the contributing researchers or anyone else who has supplied help or information.

I do not doubt that this book, like its predecessor, will cause controversy – amongst ufologists if no one else. But earth lights theory has become stronger to some extent since 1982, so hopefully there will be more eyes prepared to see, and more ears willing to hear. All theories take time to be accepted, but there really cannot any longer be reasonable doubt about the existence of these earth lights. There are now known locations, after all, where one can go and regularly witness them! Their nature and means of manifestation are, however, another matter. That is where the focus of further debate and research should be directed, and not the tiresome, time-wasting and superseded pseudo-problem of whether or not the lights actually occur. They do exist, and it is time we prepared to open up to their revelation.

PART 1
THE COMING OF THE LIGHTS

1 EARTH'S LIGHT SHOW

Our planet produces lights. A range of light phenomena emerges naturally from the Earth's own processes, but they are not fully understood. Some types of terrestrially produced lights, even though occurring fairly regularly and reported quite frequently and reliably, have barely been perceived by mainstream science or, more unexpectedly, by many of those people called 'ufologists' whose self-appointed role is to study unexplained lights and aerial anomalies. In short, we currently have a cultural blindness to a remarkable display of terrestrial phenomena that certainly have much to teach us. This blindness is one of the intellectual tragedies of our time. In some respects, people living in technologically advanced industrial societies may be less aware of some of the more intimate processes of our planet than earlier, simpler peoples; a fact which has implications for ecology and our impact on the terrestrial environment as well as for the study of unusual light events.

Lights from the Earth can be broken down into a few broad categories, and we need to look at these in order to confirm that currently inexplicable lightforms can indeed be produced by natural processes, and to track down the fugitive earth lights phenomenon itself which hides amongst these other, probably related, categories.

BALL LIGHTNING

Electrical, thundery weather seems to be associated with one class of remarkable lightform called 'ball lightning'. This can take on many colours and range from pea-size to the dimensions of a beachball or larger. Although usually round, ball lightning can also appear in more bizarre shapes such as cylinders, spheres with protrusions, and dumb-bell forms. Some forms of ball lightning are translucent or even transparent, while records also exist of cloudy, opaque forms. There are accounts, too, of totally black phenomena occurring in similar circumstances to ball lightning. The phenomenon can hover, float languidly, fall to the ground, remain stationary, or move as if purposefully.

Probably the most scientific eyewitness description of ball lightning is that recorded by R. C. Jennison in *Nature*, November 1969. As well as confirming the ability of ball lightning to occur within a metal-screened environment, which a straightforward electrical effect would not be expected to do, the account also brings up another curious factor often associated with ball lightning – the apparent lack of heat generation. Jennison was seated near the front of the passenger cabin of an all-metal airliner on a late-night flight from New York to Washington in March 1963. The aircraft encountered an electrical storm in which it was engulfed in a bright and loud electrical discharge:

> Some seconds after this a glowing sphere a little more than 20 cm in diameter emerged from the pilot's cabin and passed down the aisle of the aircraft approximately 50 cm from me . . . .
>
> The observation was remarkable for the following reasons, (i) The appearance of the phenomenon in an almost totally screened environment; (ii) the relative velocity of the ball to that of the containing aircraft was 1.5 + /-0.5 ms^{-1}, typical of most ground observations; (iii) the object seemed perfectly symmetrical in all three dimensions and had no polar or toroidal structure; (iv) it was slightly limb darkened having an almost solid appearance and indicating that it was optically thick; (v) the object did not seem to radiate heat; (vi) the optical output could be assessed as approximately 5 to 10 W and its colour was blue-white; (vii) the diameter was 22 cm + / – 2 cm, assessed by eye relative to surroundings; (viii) the height above the floor was approximately 75 cm; (ix) the course was straight down the whole central aisle of the aircraft; (x) the symmetry of the object was such that it was not possible to assess whether or not it was spinning.

While it is most witnesses' experience that ball lightning does not give off heat, there are some reports that say otherwise. These and other variations in ball lightning descriptions seem to indicate that even within this general category of the phenomenon there are a range of manifestations.

An account of a ball lightning encounter was given to me personally by the witness, William Becker, a professor and chair of industrial design at the School of Art and Design, University of Illinois at Chicago, and a researcher into solar energy equipment design. It is a remarkably detailed and close observation, and has not been previously published:

> During the summer of 1958 . . . I was with a canoeing group of High School buddies in the upper Minnesota regions north of the city of Grand Marias . . . . We finally had a bad break in the weather. . . . We spotted a deserted cabin that could be used as a campsite.
>
> . . . The six of us split up . . . and my friend and I ended up in the back room

of the cabin. The rain was now a downpour, but the cabin was still stuffy from the day's earlier heat; so after unrolling my sleeping bag I went and opened the window a crack. . . . The sky had now darkened to a nightime gray . . . we both looked up together to see what looked like a flashlight moving around outside the opened window. My immediate thought was that it was one of the pranksters in our group. . . . To our complete amazement, the 'flashlight' illumination began to squeeze through the open, one-inch crack above the window sill. As we watched, a 'bubble' of light emerged from the space in the open window and slowly floated into the room.

The lightball I estimated to be just larger than a basketball. It hung in the air a moment and seemed to be making a lot of rapid short movements which added up to overall 'smooth' motions. The ball had a bright outer perimeter of yellow-white light with an inner core of darker orange light. As it moved from the window on my right to just in front of me, I saw what looked like 'worms' or short 'strings' of light writhing at its center. It made no sound, but slowly descended towards the floor where an old black and white, Indian-style rug still lay inside the door. Inexplicably, the lightball moved down just over the rug, and as it continued across the room from my right to my left, it 'traced' or followed the dark patterns in the rug on its course. It then slowly proceeded toward the end of the room where it angled away and shrunk in size as if finding an escape hole in the corner kick boards. In a moment, it was gone.

As my friend and I turned to each other to scream our amazement at the event, a sharp, piercing report, like a loud firecracker, rang out from behind the wall of the room. . . . The next morning we inspected the outside of the cabin and found a broken drainpipe connection dug into the ground outside the cabin wall where the explosion had occurred. On the wall where the light had entered, nothing unusual or metallic seemed apparent. . . . By the end of the trip we hypothesised that lightning must have interacted with some of the many copper deposits in the area to produce the ball.

So, ball lightning is so termed because it is occasionally seen in association with thundery conditions, is assumed to be electrical in nature, yet displays coherent forms that defy current explanation and behaves in ways not associated with electrical events as presently understood. In short, it clearly possesses characteristics that elude our present understanding of physics. Many scientific papers have been written trying to explain the phenomenon with suggestions ranging from anti-matter to standing waves of electromagnetism, but none have been completely successful.

Furthermore, light balls are sometimes reported that are classified as 'clear sky' ball lighting, when no thundery conditions are local to the event. Indeed, I witnessed such a lightball myself in the summer of 1983. I was situated in a building on a hillside to the southwest of the small Welsh town of Llanfyllin, Powys, at map reference SJ 14051955, and had a view through the window to the northeast. Weather conditions were stable and clear. A solitary cloud hung close to the hilly northeast horizon less than two miles away between two hills called Grave and Jericho, at approximately SJ 160207, I saw a ball of pearly white light drop vertically but gently from the cloud to disappear in a few brief moments behind the horizon. There was no sound, nor any unusual meteorological circumstances that could be noted accompanying the event.

There is no reason to claim such 'clear-sky' events as ball lightning. The only rationale for associating a ball of light with thunder and lightning is if such electrical conditions accompany its appearance!

Ball lightning is now accepted by most scientists, though it seems there are a few who still resist admitting to its existence.

WILL-O'-THE-WISP

Another class of lights produced by Earth are the usually flame-like forms that frequent marshy ground, known variously as 'Will-o'-the-Wisp', 'Jack-o'-Lantern', 'Kitty Candlestick', 'Ignis fatuus', 'Corpse Candles' (*canwll corf* in Welsh) and by many other appellations. In folklore, they were interpreted as wandering sprites liable to lead travellers to their doom, or were considered harbingers of death, and were widely believed to hover over fresh graves. A classic account of Will-o'-the-Wisp was given by L. Blesson, a Berlin engineer, in 1833:

> The first time I saw the ignis fatuus was in a valley, in the forest of Gorbitz, in the New Mark. This valley . . . is marshy on its lower part. . . . During the day, bubbles of air were seen rising from it, and in the night blue flames were observed shooting from and playing over its surface. As I suspected that there was some connection between these flames and the bubbles of air, I marked during the day-time the place where the latter rose up most abundantly, and repaired thither during the night; to my great joy I actually observed bluish-purple flames, and did not hesitate to approach them. On reaching the spot they retired, and I pursued them in vain. . . . I conjecture that the motion of the air, on my approaching the spot, forced forward the burning gas. . . . On another day, in the twilight, I went again to the place . . . the flames became gradually visible, but redder than formerly, thus showing that they burnt also during the day. . . . I was able to singe paper [against the flame). . . . I next used a narrow slip of paper, and enjoyed the pleasure of seeing it take fire. The gas was evidently inflammable. . . . On the following evening, I went to the spot and kindled a fire on the side of the valley. . . . I hastened with a torch to the spot from which the gas bubbled up, when instantaneously a kind of

> explosion was heard, and a red light was seen over eight or nine square feet of marsh, which diminished to a small blue flame, from two and a half to three feet in height, and continued to burn with an unsteady motion.[1]

This report squares with the general (but unproven) assumption that this class of lightform is burning 'marsh gas' (largely methane) issuing from swampy ground. It contradicts most other Will-o'-the-Wisp accounts, however, where flames appearing over boggy ground are described as *not* producing heat. For example, an excellent description was provided by E. Knorr, a professor of physics at Kiev, in 1853. He saw a Will-o'-the-Wisp illuminating an area by the roadside at a point where a bridge crossed a swampy stream: 'Bushes, rushes and grass were lit up so brightly by the light that for some time I gazed at the lovely picture entranced.' But he then galvanised himself into an investigation of the phenomenon. Stretching full length across the boggy area, he was able to come within about eight inches (20 cm) of the light source – a flame about five inches (12 cm) in height, and up to an inch and a half (3.5 cm) broad. It was cylindrical in shape, and yellow in its core but appearing intense violet on its edges. The flame was steady, but moved backwards and forwards slightly when Knorr made a draught by waving his handkerchief. The physicist then held the brass ferrule of his walking stick in the flame for some 15 minutes. When he drew the stick back, the ferrule was still cool.[2]

Certainly, the closer the nature of the phenomenon is studied, the more problematical it becomes. This was recognised by researcher Phil Reeder, writing in 1986:

> It is a well-known fact that rotting organic matter produces methane (CH_4), a low order hydrocarbon. However, methane does not combust spontaneously: the current use of natural gas would not be feasible if methane, its chief constituent, was so unstable. A suggested trigger for combustion is liquid hydrogen phosphide (P_2H_4), traces of which can sometimes be found where organic matter is decaying. ... Assuming that hydrogen phosphide is the trigger, do the characteristics of burning methane match the observed phenomena? Methane rising to a point on the surface at a continuous and regulated rate if ignited would appear as a flickering and fixed flame, like a Bunsen burner ... [this] may well account for some sightings, yet the phenomenon is often seen moving through the air, even against prevailing winds.[3]

Allan A. Mills, a lecturer in planetary science, has commented that 'explanations of the Will-o'-the-Wisp have been as numerous as they are unsatisfactory'.[4] He carried out preliminary experiments but was unable to replicate the phenomenon as reported, and also felt that the suggestion of the phosphine or phosphorous ignition trigger could be a 'false trail'.

It is clear that the chemistry of Will-o'-the-Wisp lights is incompletely understood, and at best can only explain a very strict form of marsh light event. Study soon reveals that lightforms that cannot possibly be explained as burning methane or other gases have been loosely lumped under the heading of 'Will-o'-the-Wisp' or related names by witnesses seeking some kind of label. This was recognised by a Mr Andrew Lang writing in the *Illustrated London News* around the turn of the century:

> I am well acquainted with the set of lights which are often seen by the people of Ballachulish and Glencoe villages on the south side of the salt-water Loch Leven. . . . They are bright lights which disport themselves on the north side of the loch, where steep hills descend to the level, and to the road along the level, leading to the head of the loch. They rush, as it were, along the road, then up the hill, then to the water edge, and so on, and are visible not only to the Celtic natives but to the English tourist. The ground is not marshy, even on the level, and the phenomena, though doubtless natural and normal, have not yet found a scientific explanation. They are not what people call 'corpse candles' . . . .

A letter from C. Leeson Prince, published in *Notes and Queries* in 1891 shows the common confusion in earlier accounts of unusual light phenomena with Will-o'-the-Wisp terminology:

> I wonder whether you have ever observed the Will-o'-the-Wisp which for several years we have observed from the windows of the house here [Crowborough, Sussex] facing . . . in the direction of Gill's Lap. He is a stately fellow, and does not condescend to dance, hopping and skipping close to the ground, like some of his brethren, but prefers a sort of stately minuet high up in the air above the tree tops. He was magnificent the night before last, and I never saw him so high. His appearance always betokens bad weather, and the higher he goes, the worse the weather.[5]

This simply could not be the sort of presumed chemiluminescent effect claimed for Will-o'-the-Wisp phenomena. The account obviously suggests meteorological associations, yet does not readily fall into the category of ball lightning either. Even many of the flame-like lights which have been reported cannot easily be identified as Will-o'-the-Wisps. For example:

> In 1888 I saw a light which must have been a 'will-o'-the-wisp'. It was in Florida and the day was misty and damp, and I first noticed it coming across the lake upon which my property stood. After continuing an irregular course it made for the railway. I pursued the light, which was large and flickering, for half a mile and apparently got within 100 yards of it when it suddenly collapsed or disappeared, I cannot say which.[6]

The range of surfaces and distance covered by this large light put it beyond the scope of the suggested chemiluminescent explanations of Will-o'-the-Wisp phenomena. Many of the accounts lodged under the 'Will-o'-the-Wisp' category are, in fact, remarkably similar to some UFO reports, as in the case from Sussex quoted above. It is perhaps significant, as Phil Reeder points out, that 'while sightings of Will-o'-the-Wisps have decreased towards the end of last century, sightings of UFOs have increased'. In other words, one of the catch-all terms for strange light phenomena may have been 'Will-o'-the-Wisp' in Victorian times, and has now become 'UFO' – or, indeed, 'ball lightning'.

EARTHQUAKE LIGHTS

A third broad category of terrestrial light phenomena is that of earthquake lights (EQLs). These come in a wide range of forms – such as streamers and aurorae-like displays across the sky, balls of light, glows in the atmosphere, sparkles of light on hillsides, 'slow' lightning – and occur in association with some, but by no means all, earthquakes. They can appear before, during and after quake activity, and sometimes at distances of tens of miles from the epicentre of such seismicity.[7]

Lights have been reported down the generations by victims of earthquakes worldwide, but were shunned by most modern scientists because of the anecdotal nature of the evidence. The dawn of scientific awareness of EQLs seems to have been in 1910 when the Italian scientist I. Galli published a report on 148 observations of light phenomena made in Italy during the nineteenth century. But that such lights remained broadly outside the pale of scientific acceptability was confirmed by the books of Charles Fort, an American who specialised in collecting material 'damned' by official science of the early decades of the twentieth century. Fort was (typically) perceptive enough to note the connection between quake activity and the appearance of anomalous lights. In his *New Lands* (1923) for instance, Fort wrote about the nineteenth-century earthquake activity around Hereford, near the Welsh border, and referred to the associated incidence of light phenomena:

> At 9.30 p.m., Jan. 25, 1894, at Llanthomas and Clifford, towns less than 20 miles west of Hereford, a brilliant light was seen in the sky, an explosion was heard, and a quake was felt. Half an hour later . . . at Stokesay Vicarage, Shropshire, was seen . . . an illumination so brilliant that for half a minute everything was almost as visible as by daylight.
>
> . . . 'a strange meteoric light' . . . was seen in the sky, at Worcester, during the quake of Dec. 17, 1896 . . . in another town, 'a great blaze' was seen in the sky. . . . In an appendix to his book *The Herefordshire Earthquake of 1896*, Dr Charles Davison says that at the time of the quake (5.30 a.m.) there was a luminous object in the sky, and that it traversed a large part of the disturbed

> area. He says that it was a meteor, and an extraordinary meteor that lighted up the ground so that one could have picked up a pin. With the data so far considered, almost anyone would think that of course an object had exploded in the sky, shaking the earth underneath. Dr Davison does not say this. He says that the meteor only happened to appear over a part of this earth where an earthquake was occurring, 'by a strange coincidence'.
>
> Suppose that, with ordinary common sense, he had not lugged in his 'strange coincidence' . . . .

In the Idu Peninsula earthquake of November 1930, researchers Terada and Musya collected some 1500 EQL reports. Examples of this carefully observed information began to infiltrate the scientific literature during the 1930s. Luminous phenomena were also reported to Musya from four quakes occurring in the year following the Idu Peninsula event. This growing body of data was enhanced by sequence photographs of EQLs taken during the Matsushiro earthquake swarm of 1965–7 (see picture on back jacket). It is to be assumed that most seismologists today accept the reality of EQLs, as by 1973 John S. Derr, a seismologist with the US Geological Survey, could write: 'The existence of luminous phenomena, or earthquake lights, is well established.'

One of the greatest earthquakes known of in relatively modern times was the New Madrid Earthquake of 1811 in America. It was in fact a succession of shocks in the central Mississippi valley affecting southeastern Missouri, northeastern Arkansas, and western Kentucky and Tennessee. 'Beginning on December 16, and lasting more than a year,' says a report of 1912, 'these shocks have not been surpassed or even equalled for number, continuance of disturbance, area affected and severity by the more recent and better-known shocks at Charleston and San Francisco. As the region was almost unsettled at that time relatively little attention was paid to the phenomenon, the published accounts being few in number and incomplete in details. For these reasons, although scientific literature in this country and in Europe has given it a place among the great earthquakes of the world, the memory of it has lapsed from the public mind.' Because of the paucity of records, many light phenomena associated with this huge quake, which produced marked earth disturbance over an area of around 40,000 square miles and registered tremors over half the entire USA, must have escaped documentation. Nevertheless, the 1912 report does mention some examples:

> The phenomena of what may be termed 'light flashes' and 'glows' seem so improbable that they would be dismissed from consideration but for the considerable number of localities from which they were reported. Dillard, in speaking of the shocks (not especially the first one), says: 'There issued no burning flames, but flashes such as would result from an explosion of gas, or from passing of electricity from cloud to cloud.' Lewis F. Linn, United States

> Senator . . . says the shock was accompanied 'ever and anon [by] flashes of electricity . . . '. Another evidently somewhat excited observer near New Madrid thought he saw 'many sparks of fire emitted from the earth'. At St Louis gleams and flashes of light were frequently visible around the horizon in different directions, generally ascending from the earth. In Livingstone County . . . the atmosphere previous to the shock of February 8 was remarkably luminous, objects being visible for considerable distances, although there was no moon. 'On this occasion the brightness was general, and did not proceed from any point or spot in the heavens. It was broad and expanded, reaching from the zenith on every side toward the horizon. It exhibited no flashes nor coruscations, but, as long as it lasted, was a diffused illumination of the atmosphere on all sides.' At Bardstown there are reported to have been 'frequent lights during the commotions.' At Knoxville, Tenn., at the end of the first shock, 'two flashes of light, at intervals of about a minute, very much like distant lightning,' were observed. Farther east, in North Carolina, there were reported 'three large extraordinary fires in the air; one appeared in an easterly direction, one in the north, and one in the south. Their continuance was several hours; their size as large as a house on fire; the motion of the blaze was quite visible, but no sparks appeared.' At Savannah, Ga., the first shock is said to have been preceded by a flash of light.[8]

The report at least pays lip service to a certain scepticism in keeping with the orthodox outlook of the time regarding EQLs, but admits that it is 'improbable' that the lights associated with the New Madrid events were 'entirely imaginary'. It goes on to make a further interesting observation:

> Bearing on the origin of the flashes or glows the observations of several of the captains of ocean liners in the Tropics at the time of the recent severe disturbance in Mexico (1907) are of significance. They reported that on the night on which they afterwards learned that the earthquake had occurred strong glows in the sky, resembling the auroras of northern latitudes, were seen. As these were not reported father north the view suggests itself that they were due to magnetic disturbances depending upon or related to the severe earth disturbances going on at the time. It is not improbable that similar magnetic manifestations were associated with the New Madrid shock.

An item in *Nature* of 1872 reported that immediately following the great shock of a Californian earthquake recent to 1872, two witnesses observed 'sheets of flame on the rocky sides of the Inuyo Mountains' about half a mile from the Eclipse Mines. 'These flames, observed in several places,' the report continued, 'waved to and fro, apparently clear of the ground, like vast torches. They continued for only a few minutes.'

An Italian earthquake researcher called Benevilli reported during the 1786 Ligurian earthquake that while the tremors were occurring 'one felt electrified or magnetized and in several places "sulfur odors" appeared as well as fires moving across the earth. Falling stars and flashing sparks flew through the atmosphere."[9]

The 1906 earthquake around Valparaiso, Chile, yielded a remarkable range of EQL reports. A policeman at San Bernado saw 'electrical discharges toward the southeast which emitted an intense red light'. 'Flames in the form of snakes' were also seen. Captain Rafael Gonzales in Limanche Viejo observed fiery lights of zig-zag shape moving close to the ground, and other spectators saw red-violet patches in the sky during the earthquake. During the minutes of the actual earthquake, meteorologists at the Santiago observatory saw 'enormous electrical discharges that appeared to cover the entire northeast horizon from the tree tops to an altitude of about 30 degrees'. A school principal at Rancagua reported that 'when the third shock came, great flames of electric blue color could be seen in the west which rose until they lost themselves in the clouds.'

EQLs are sometimes reported that seem very similar to many UFO and ball lightning accounts. Just before a quake in 1887 in Liguria, villagers in Loano saw a red light above the village that 'resembled a flame'. Pilots of two ships off the Chile coast during the Valparaiso earthquake saw 'seven or eight balls rising from a house' that was burning high up on some hills. The balls moved a distance to the east then dropped, with detonations marking the fall of each one. During the same quake, members of one family saw fireballs as big 'as the moon'. One was stationary and was 'giving off smaller balls with tails'. During the 1930 Idu Peninsula earthquake a 'straight row of round masses of light' was observed in the sky to the southwest of the epicentre. A similar sight was observed prior to the 1957 Charnwood quake, in Leicestershire, England: there were numerous reports of 'tadpole-shaped lights' flying in rows high in the sky.

Lights are seen above and within the sea, too. In *When the Snakes Awake* (1982) Helmut Tributsch observes:

> . . . unexplained . . . glowing lights at sea may be traceable to submarine quakes. Of 70 sea captains' reports of light appearances at sea which were evaluated by Kurt Kalle of the German Hydrographic Institute in Hamburg, 19 described a kind of light-ball bombardment from the ocean depths . . . glowing balls, about a metre across, rose up from the depths and burst soundlessly on the ocean surface, turning into garish discs of light about 100 metres (328 feet) in diameter and then quickly going out. These phenomena have been reported only from volcanically active tropical ocean regions.

Other extraordinary light phenomena are reliably reported by seafarers – especially in the Indian Ocean. Great wheels and bands of light are periodically seen fanning soundlessly across the ocean. These usually seem to be on the surface of the water, but are sometimes reported several feet above. These are amongst the most mysterious light phenomena reported anywhere on the planet, but absolutely nothing is known about them. They may be associated with tectonic events, like EQLs, but, if so, the form of the light defies any known geophysics. They may simply be another relative in Earth's family of natural light phenomena. An intriguing observation was made in 1951, however, when Captain F. G. Baker noted that at the moment his ship's radar was switched on 'most brilliant boomerang-shaped arcs of phosphorescent light appeared in the sea, gyrating in a clockwise direction to starboard and anti-clockwise to port, but all sweeping inwards towards the ship from points situated from five to six points on either bow and some two miles distant.' The whole ship was illuminated as if floodlit. The captain wondered if the ultra-high frequency emissions from the ship had somehow had an effect 'upon some submarine substance.'[10]

EQLs conclusively show that geological factors such as seismicity can play a part in producing a multitude of lightforms in the atmosphere, and possibly in the ocean; lightforms that can move, stop, fly in formation, change shape, occur on the ground, close to the ground and high in the air. It is obvious that they have characteristics almost identical to those described in UFO reports. The only reason to distinguish them from UFOs is their association with tectonic activity - usually a severe earthquake. However, EQLs have also been reported in association with quite mild tremors.

Some EQLs are reminiscent of ball lightning descriptions. Perhaps tectonic activity can interact with electrical properties in the atmosphere causing ionisation glows. It is a common belief, if scientifically elusive, that meteorological changes accompany the onset of earthquakes - so-called 'earthquake weather'. This suggests some form of geological-meteorological link.

A number of theories put forward to explain EQLs have included the *piezo-electric effect*. This involves the transduction of mechanical energy into electrical energy and *vice versa*: when a crystal has pressure put on it, electrical charges occur across its surfaces; when varying voltages are applied to a crystal, it expands and contracts - a mechanism made use of in quartz watches, for example. It is easy to see that enormous pressures in the earth at times of tectonic unrest, with huge bodies of crystal-bearing rocks crushing against one another, could easily produce enormous discharges of electrical energy. This would occur particularly along geological fault lines - fractures in the earth's crust caused by geological, tectonic, action - where movement could occur. It is estimated that 10,000 to 100,000 volts per square metre could be produced in such circumstances, depending on pressures involved and the piezo-electric efficiency of the rocks subjected to such forces. But while this might produce lightning-like effects in some instances, it hardly seems capable, on its own, of accounting for relatively stable lightforms such as spheres high in the air. Other theories for EQLs, such as the violent low-level oscillation of air molecules or the emission of ignited gases from fissures in the ground during earthquakes, similarly fail to explain all the forms of the EQLs that have been reported from around the world. Science still has to explain how specific shapes of light can keep their forms intact and manoeuvre in the air.

A phenomenon which may be associated with EQLs is *mountain peak discharge*. It has been noticed for a great many years that mountain tops can produce glowing 'auras', rays and beams of light, and even emit balls of light.[11] These mountain peak lights have been noted especially in the Andes, though they have been observed throughout the world at suitable locations. Their intensity seems to increase during earthquakes. We may have here a type of light event that draws on both geological and atmospheric electrical mechanisms. Some hills as well as mountains seem able to produce this type of light emission, and certain examples of both seem particularly haunted by lights. Lights are produced in such cases that simply cannot be explained by known physics. It is an aspect of earth lights that we return to in Chapter 6.

'Volcanic lightning' is another type of light emission certainly associated with geological – tectonic – activity, and must be a close relative of EQLs and possibly mountain peak discharge also. In 1902, for example, a paper in *Nature* reported extraordinary light effects radiating out of an erupting volcano, Mont Pelee: 'Zigzagging and flickering flashes alternating with, or being accompanied by, reddish globes, which ascended and exploded, and shot out stars and long rays.' A point some 40 miles (64 km) from the volcano also displayed another 'large focus of electric energy' which produced similar, though less extensive, light effects to Mont Pelee, and seemingly in 'distinct relation' to the effects emerging from the volcano. This secondary centre of luminous activity was accompanied by 'curious glowing globes, which burst and shot out tongues of lightning'. Again, three days into an eruption of Mount Vesuvius in 1906, Frank A. Perret, observed 'a thin, luminous arc' flashing out of the crater immediately prior to each explosive eruption from the volcano. Around the end of 1986 and January 1987 volcanoes were erupting in Cameroon, Africa, around Lake Nios. Over 1700 people were killed after an enormous cloud of poisonous gas had been released by the volcanic activity. On 7 January 1987, the French volcanologist M. Haroun Tazieff reported 'three explosions, three violent detonations' around Lake Nios in a five-minute period. One of these 'was accompanied by strong flashes of light', Tazief observed.[12]

EARTH LIGHTS

This brief and simplified survey of lights produced by the Earth demonstrates a few key factors. The most crucial is the *certainty* that the Earth produces a range of lightforms by natural processes. That is a major psychological threshold, a Great Barrier Reef of disbelief, that many people have to overcome. The processes are poorly understood. Even modern science does not have a firm grasp of all the workings of land, sky and sea. The hard lesson is that we do not know our planet as well as we would like to believe. It seems clear that the vast range of mechanisms that can be marshalled in various ways by the Earth produce energy effects that at least bend the rules of chemistry and physics as we know them, and in certain circumstances might even be presenting us with unguessed forms of electromagnetism, or as yet unknown energies or energy configurations. The Earth draws on mightier resources and realities than any laboratory, and nature teaches science, not the reverse.

There is something more. In this chapter we have identified the three main conditions for the appearance of anomalous lights: exceptional electrical meteorological conditions, unusual emissions of chemiluminescence, and particular instances of geological disturbance. Theories applied to each of these categories of light are only effective for very strict versions of luminescence involved, and then in rather unsatisfactory or limited fashion. It is obvious in scrutinising the record that lightforms are reported under the headings of all three categories that cannot be properly ascribed to them nor explained by the theories available. They thus constitute *another* class of lightforms. For simplicity, we may call these fugitive phenomena *earth lights*.

Ball lightning, Will-o'-the-Wisp and EQLs are each a spectrum of phenomena within themselves, yet perhaps share some geophysical common denominators. Earth lights may likewise draw on a mix of electrical, geological and gaseous conditions, but their essential characteristics of flexible manoeuverability, range of relatively stable, regular forms, and tendency to haunt specific locales suggest that other, presently unknown or unidentified factors are also involved. Examples of earth lights have been briefly included in some of the descriptions above – for example the lights at Loch Leven – and many more will be given in following chapters. We will see that the most reliable strand of information we can so far glean about these enigmatic lights is that they often display a relationship with certain *geological* features and events, and that seems the road down which their study needs to proceed. But for now, it is enough that we *perceive* that there is this further category of light phenomena springing spontaneously from the body of Mother Earth.

But just as earth lights begin to be separated out from the camouflage of possibly related but somewhat different light phenomena, they become swallowed up by that voracious and ill-disciplined monster, the UFO. It is a monster that has to be challenged.

2 THE CASCADES CATALYST

'Unidentified Flying Objects' ('UFOs') is the usual term employed to describe strange or anomalous aerial phenomena. Misidentifications of planes, planets, moon, weather balloons, satellites and so on are to blame for most UFO reports. Others are occasioned by abnormal atmospheric conditions creating distorted 'mirage' images of mundane objects such as aircraft or astronomical bodies. Other UFO reports are simple lies or hoaxes, while a further percentage results from complex and little-understood psychological processes taking place within the 'witness'. Thus the great majority of UFO reports can be furnished with explanations. But no one who has taken the time and effort to study the UFO enigma in any depth can reasonably believe they account for them all.

'Ufology' is the unofficial study area involved with the investigation and evaluation of these matters. Many researchers, nearly all part-time, study aspects of the enigma in many countries of the world. Because such work is unfunded and amateur, the quality of research is uneven, and resources limited. Serious UFO study is hampered by the media, which almost always wants to treat the subject in a jocular or sensationalist manner. It is extremely difficult to conduct a sane, balanced discussion about UFOs in public – the Press want to know only about aliens and extra-terrestrial craft, while scientific sceptics, largely because of the media circus, consider the whole matter as belonging to the lunatic fringe.

In addition to these various difficulties, ufology itself creates its own chaos. It is a volatile mixing of many different types of mentality drawn to the UFO mystery for a variety of reasons and psychological drives. The truth is that UFOs and ufology are different things, and come into contact only occasionally. Modern ufology encompasses a spectrum of 'explanations' for UFOs, ranging from extra-terrestrial spacecraft on the one hand, to UFOs as complex psychological and sociological ('psychosocial') manifestations, without any objective, external reality, on the other. The earth lights perception lies somewhere in the middle of this range of opinion.

Ufology is thus the penumbra surrounding the reported occurrence of UFOs. But where did it come from; when did it start? Strange aerial phenomena have been reported throughout history, but they came to be considered collectively as 'flying saucers', UFOs, alien spacecraft only recently. The catalyst was a fairly unexceptional sighting made by Kenneth Arnold while flying over the Cascade Mountains in Washington State, northwest USA, in 1947. But though it was a run-of-the-mill sighting, it took place in exceptional circumstances, and I make no apologies for reviewing this celebrated case here. Although old, there is still fresh information to be gleaned from Arnold's account, I feel, by looking carefully at three aspects: the terminology used by Arnold in his account, the sociological context in which his experience occurred, and the information the physical environment of the reported sighting can give us. These aspects will be studied at various points in this chapter, but first, we have to look at Arnold's report itself. I draw directly on his own accounts as given in 1952[1] and 1977[2], as well as from his 1947 report presented to the authorities shortly after the incident.[3]

A CASCADE OF FLYING SAUCERS

On the clear, sunny afternoon on 24 June 1947, experienced pilot Kenneth Arnold was flying east across the Cascade mountain range. He was flying at about 9200 feet (2814 metres) when suddenly 'a tremendous flash appeared in the sky. It lit up my whole aircraft, even the cockpit, and I was startled.' He thought he was in a near collision with an aircraft which had flashed past him reflecting sunlight off its wing surfaces. He looked all around, but saw nothing except a distant plane. Then the flash occurred again:

> This very bright flash, almost like an arc light, was coming from a group of objects far up to the north of Mount Rainier in the area of Mount Baker, which is almost in line with Mount Rainier and Mount Adams. I saw a chain of peculiar aircraft approaching Mount Rainier very rapidly . . . They seemed to fly in an echelon formation. However, in looking at them against the sky and against the snow of Mount Rainier as they approached, I just couldn't discern any tails on them, and I had never seen an aircraft without a tail! These were fairly large-sized and there were nine of them.[4]

Arnold 'assumed these were military craft', yet was puzzled by their echelon formation which was not standard for US military conventional formation, nor for any other nation he knew of. The objects were in a diagonal chain-like formation, but each object was independent and moving with its own rhythm within the overall formation. The manner in which 'these craft . . . fluttered and sailed, tipping their wings alternately and emitting those very bright blue-white flashes from their surfaces' fascinated

Arnold. 'At the time I did not get the impression that these flashes were emitted by them, but rather that it was the sun's reflection from the extremely highly polished surface of their wings.' Arnold recalled in 1952.

The formation was about 100 miles (160 km) away when Arnold first sighted it in the crystal clear conditions of the day, but both it and Arnold were approaching Mount Rainier at approximate right angles to each other, and the flight path of the glittering shapes eventually crossed within 23 miles (37 km) of Arnold's Callair aircraft at about the same altitude. The objects passed behind a sharp projection on Mount Rainier and, as Arnold knew where he was in relation to the mountain, he was able to make a reasonably accurate estimate of the distance. As the formation passed Mount Rainier, Arnold took the opportunity to clock the time it took to reach Mount Adams about 50 miles (80 km) to the south. He did not work out the figures there and then, but he knew they were flying very fast, 'faster than our P-51s and any military plane I knew of'.

Between Mount Rainier and Mount Adams is a plateau called Goat Ridge. This is 5 miles (8 km) long and by this Arnold was able to estimate that the formation of objects was a similar length. They hugged the mountainous terrain, weaving in and out of peaks between Rainier and Adams. Arnold judged the diameter of each object to be about 100 feet (30 metres). 'When they gave off this flash,' he said, 'they appeared to be completely round. When they turned lengthwise or flatwise to me, they were very thin.' (Figure 1.) He went on:

> Now, as they flew past Goat Ridge the second from the last one seemed to turn its rear end toward me. I guess that's the best expression I can use, and I could see that it wasn't round at all. I got the impression they were rather like tadpoles. They had a little sort of peak at the center of their rear trailing edges. But I couldn't be positive they were all of the same design as the one I particularly noticed, or whether that one was a little larger. . . . I couldn't see the last ship too well because it was fluttering and jerking very rapidly. From the way they performed, I thought if there were human beings in them, they would have been made into hamburger.[5]

Arnold had an 'eerie feeling', and made for Yakima, the same general direction in which his mysterious discs had been travelling. At Yakima airfield, Arnold immediately told other fliers what had happened. One of these suggested that Arnold must have witnessed 'some of those guided missiles from Moses Lake'. Although he had never heard of a missile base at the lake, Arnold agreed at the time that that was probably the answer. He refuelled his plane and flew on to Pendleton, Oregon, as one leg on his way back to his home state of Idaho. On the way, he made some rough calculations of the speed of the formation from the timing he had done near Mount Rainier. He came up with an answer of around 1700 mph (2737 kph) – an unheard-of speed for aircraft at that time. News of Arnold's sighting travel-

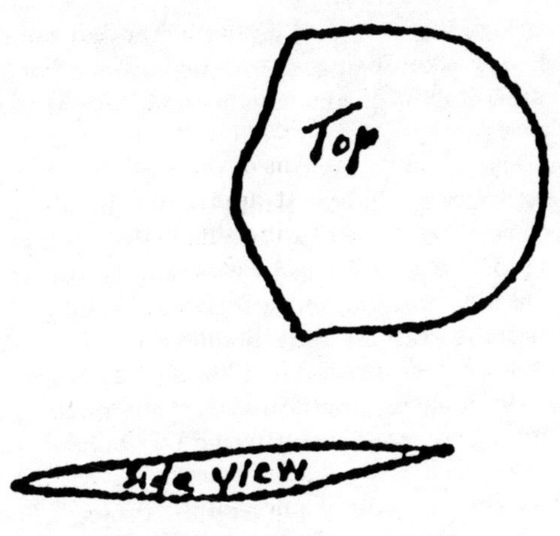

FIGURE 1 Kenneth Arnold's own sketches of the flying discs he saw.

led ahead of him, and when he landed at Pendleton, there was a group of people waiting for him. 'When I got out of my plane no one said anything,' he recalled. 'They just stood around and looked at me . . . but before very long it seemed everybody around the airfield was listening to the story of my experience.' Convinced he had got his calculations about the speed of the object wrong, Arnold and others at Pendleton repeatedly worked and reworked the figures. Making every allowance possible, the speed still came out at over 1300 mph (2093 kph). Arnold went to report his sighting at the FBI office in Pendleton, but it was closed. He then talked to journalists at the *East Oregonian*. One of them, Bill Becquette, asked how the objects flew. 'Well, they flew erratic,' Arnold replied, 'like a saucer if you skip it across water.' Becquette misinterpreted this statement, describing the objects themselves as 'saucer-like' in his Associated Press despatch. This is where the popular usage of the term 'flying saucers' originated, though it was not the first time that it had been used to describe an unexplained aerial phenomenon: in 1878 farmer John Martin of Dallas, Texas, described seeing a large orange shape in the sky. 'When it was directly overhead,' Martin stated, 'it was the size of a large saucer and evidently at a great height.[6]

Arnold's story went out on the news wires, and that night he was besieged by reporters and press agencies 'of every conceivable description'. Letters, telegrams and phonecalls flooded in to him. After being pinned down at Pendleton for three days, Arnold could take no more and flew back to his home in Boise, Idaho. There he met up with Dave Johnson, aviation editor of the *Idaho Statesman*. To Arnold's concern, it soon became clear that this informed man did not think that the Cascades objects could be examples of any new device developed by the United States. For the first time, Kenneth Arnold began to feel he had witnessed something totally inexplicable. He was soon being questioned by military intelligence officers, and became convinced that neither they nor anyone else knew what the things were.

We now know that Arnold's sighting came early in a 'wave' of similar observations by witnesses around the United States. In September 1947 the Air Force set up a study of such reports called 'Project Sign', and astronomer J. Allen Hynek was called in to evaluate the reports held by the Project. This evaluation programme was to go through various incarnations before it became 'Project Blue Book', the official study of UFO reports which was finally terminated in 1969. Hynek was associated with the Sign and Blue Book stages, but ultimately 'went public' as a champion of there being some sort of basic reality to the UFO phenomenon until his death in 1986.

Arnold's report (which was not the first on the Project's files) was passed to Hynek. He made a number of observations about the Idaho pilot's account. He pointed out that a sunlight reflection from objects so many miles away would hardly have captured the attention in the way Arnold described. This is a more important point than it may seem, and I return to it later. But Hynek also questioned the distance at which Arnold saw the objects, ignoring the fact that Arnold was clearly able to establish his closest distance to the objects because they passed an outcrop on Mount Rainier.

Hynek finally came to the conclusion that Arnold's discs were closer than he thought, and thus their size smaller and their speed slower than had been calculated, bringing it down to within the performance range of aircraft of the period. It has recently been pointed out,[7] however, that Hynek misread some of Arnold's statement and so his evaluation is to an extent flawed. No one, in fact, ever satisfactorily 'explained away' the sighting. It should also be noted that it was reported that a prospector claimed to have witnessed several discs over the Cascade Mountains at about the time of Arnold's sighting. The discs caused the man's compass needle to go 'wild',[8] Arnold always insisted on the truth of his Cascades sighting – and none of the investigators at the time ever doubted his sincerity – and he spent much of his spare time and money investigating other UFO reports (having further sightings), until his death in 1984.

It was Arnold's case that lit the fuse of the public imagination, and opened the floodgates for flying saucer reports in the media. It acted as a trigger for making the ages-old phenomenon of strange things in the sky a matter of open, public concern and interest. The Cascades catalyst was aided and abetted by the fact that a wave of aerial phenomena *was* being reported, so the media immediately had a 'live' subject to go at. The widespread fascination with flying saucers soon led to crops of books and movies in which flying saucers figured prominently. Within a few years people were claiming that they had gone for rides in flying saucers; UFO research groups came into existence (and so did UFO cults), and specialist journals devoted to UFO research began to be published.

Writer Hilary Evans has asked a series of questions about the Arnold case:

> Maybe the question, 'Did Kenneth Arnold see a UFO' is essentially meaningless.
>
> Maybe it would be more meaningful to ask 'Why did it happen to Kenneth Arnold and not to someone else?' or 'Why in 1947 and not some other year?' or 'Why in the United States and not somewhere else?' or 'Why did this sighting, whatever it was, make such an impression both on him and on society?'[9]

Why the sighting made an impression on Arnold is simple enough: if it happened, he was more than entitled to be impressed! Also, we know perfectly well that similar experiences to Arnold's *did* reportedly happen to other people, both before and after his 24 June sighting. If Arnold's report was not the catalyst, it would almost certainly have been one of the others a little sooner or slightly later than his, because conditions were such that the catalyst was bound to occur around the time it did, and in the country that it did. Arnold was in the optimum geographical location to witness his flying discs, as we shall see, and in a more general sense, he was in the right place and at the right time for his account of them to be taken up so swiftly and broadcast on such a scale with such effect – post-Second World War America.

A TIME AND PLACE

In an important sense *we all saw Kenneth Arnold's flying saucers*. It was the first *mass media* UFO sighting. America had the resources, the technology and the national temperament for exploiting the huge advances in all kinds of technology, including communications technology, made possible by the exigencies of the Second World War. While a shattered Europe started the slow process of repairing itself, and the Russian bear toiled in secret, America emerged from the war as the most powerful nation on Earth, undamaged physically, brash. Everywhere in the West there was the hope of a brave new technological future, but that future was closest to realisation in America. Old social orders had been dashed virtually everywhere, and everyone wanted a slice of the cake. Science was the new god, Germany had paved the way to the stars with its fearsome V-weaponry. Peenemunde, the Nazi rocket base, was plundered for its scientists and rocket engineers by both Russia and America: the revolutionary development of aerospace technology we see today was initiated immediately after the war. Space travel was to become a reality. In America particularly, this was foreshadowed by an increase in published science fiction, which in the late 1940s and the 1950s became a major *genre*. Before Arnold's sighting there was already a strongly growing sci-fi awareness through the agency of big-selling magazines like *Amazing Stories*, whose editor, Raymond Palmer, was later to persuade Kenneth Arnold to co-write a book with him[10] based on the Cascades event. The sky was the limit both metaphorically and technologically. All the old certainties had died in the war, a new world had to be built, a world fit for heroes, and heroes' children.

So the technically mature media machine existed, the background of space and science fantasy had primed the motor of the American psyche, and technology was bursting out everywhere. The scene was set. All it needed was a UFO wave, which duly came along in 1947. UFOs – 'foo fighters' (see below) – had been repeatedly seen by aircrews during the war, but the world was otherwise engaged. The 1947 wave was not even the first since the war, but it was the first in America. It was only a minor matter of months for one report or other actively to engage with the cogs of the media, the pseudopodia of the American consciousness. It happened to be Arnold's account. Becquette's conjuring of the term 'saucer-like' gave the mass media what it needed – a simple handle.

But Marshal McLuhan's 'global village' was already beginning to happen, and Arnold's story leapt oceans. Arnold himself was staggered: 'Before the night was over I had long-distance calls from London, England, from religious groups, from people who thought the end of the world was coming!' *Before the night was over*. What Arnold had seen in the depths of the Cascade mountains was flashed around the world before the next day came. This run-of-the-mill sighting had taken on special dimensions like no other before it. 'In total,' Arnold later estimated, 'I received something like ten thousand letters from all over the world. So many people came to visit me that for almost three years our home was like Grand Central Station.'

The mass preparedness to accept the idea of some kind of advanced technology in the sky had arrived; it could be conceptualised by a whole culture in a way not possible before. Arnold represented that culture.For days and weeks after his sighting, Arnold felt that he had seen terrestrial machines of some advanced kind. But the time had come when travel in space, our own let alone that of alien beings, had become a discernible reality. As it became clear that what was being seen in the sky was not likely to be the product of terrestrial technology, the door was open to extra-terrestrial interpretation. Almost imperceptibly Arnold's 'missiles from Moses Lake' became spacecraft from outer space. In a short time, the idea was simply assumed: if flying saucers where real, then they had to come from outer space. The same technological consciousness that was then initiating the greatest cycle of human exploitation of planetary resources and indifference to the natural environment ever seen, was the one that designated the nature of the flying discs as alien machines. It was an interpretation that came out of an *Earth blind* cultural context.

This process was tied up with something else. The Second World War was unlike anything that had ever been. Global war had become a reality. Everyone was at risk, at least theoretically, in such a conflict. Technology underwent unprecedented development in the war years. Nuclear power had been released in the most awesome weaponry, and threw its mushroom-shaped shadow down the post-war years. And Russia, with its alien system of Communism, was perceived as a rising, inimical superpower: an Iron Curtain was conceptualised as existing between it and the West.

The consciousness of the West in the post-war years was in a state of trauma. It lived on the edge of abysses containing possible terrors which superimposed themselves on the recent memories of war, when inhumanity had reached depths beyond nightmares. The brave new world of technology was a desperate veneer over a seething cultural paranoia. Science was to elevate humanity out of its dark, organic fears to the clean, bright light of a technological dawn. America was the pressure-cooker of the Western culture, and it was there where the extremes, positive and negative, were to show. In the late 1940s and into the next decade, the fear of Communism and of external infiltration or invasion was endemic in America. The CIA was formed. McCarthyism flowered. The need for military superiority and security was paramount. Reds might be under the beds of the West.

Arnold's experience was telegraphed into this nascent psychosocial cauldron. Machines in the sky. Russia. Extra terrestrials. Invasion. An extra-terrestrial technological devil to balance the terrestrial scientific god. Fear of invasion from the Red Planet Mars (or any other planet would do) and Reds under the Beds came from the same psychological matrix. They were interchangeable motifs. Nigel Watson noticed this in his study of UFOs and the cinema:

> The most obvious use of the contemporary interest in strange flying objects is given away by the title of *The Flying Saucer* (Mikel Conrad; 1950). Whereas in later science fiction films the aliens may be seen as a metaphor for the communists, *The Flying Saucer* has no need for such subtleties. It clearly shows the Russians stealing a saucer from a scientist in Alaska. Fortunately for the free-world and the continued production of Ma's apple pie the plans of the Russians are thwarted when the craft explodes in mid-air and the scientist gives his secrets to the U.S. Government.[11]

Film posters and book covers of the years immediately following Arnold's sighting show frightened humanity running in the streets, looking up in desperation at flying disks appearing like the eyeballs of some baleful demon in the sky.

But, as the Western psyche climbed slowly out of its post-war trauma, the extra-terrestrial motif took on softer lines. People like George Adamski emerged, claiming to have had contact with extra-terrestrial beings. These beings were humanoid Venusians, long-haired, beautiful. They warned of the dangers of nuclear war. UFO groups (or cults depending on one's viewpoint) such as the Aetherius Society appeared, according to which the Master Jesus was also located on Venus. He could beam his love down to Earth from an orbiting spacecraft. A bizarre hybrid of religion and technology had grown out of the post-war years, and the Martian archetype became eclipsed by that of Venus.

Though the world had woken up to the rumour of things seen in the sky on an unprecedented scale, people saw UFOs only out of dream-laden or nightmare-haunted eyes.

FRAMES OF REFERENCE

What had happened was that an ancient phenomenon had been given a mid-twentieth-century context. It was perceived through the cultural filters of the period. This is what has always happened to UFOs, or the many other names they have been given. The following is merely a brief 'snapshot' history of UFOs, and the instances mentioned could be replaced by countless others, but it is enough to show that the present perception of 'UFOs' is only another manifestation of a phenomenon as old as time. It provides a sketch outline of the overall context into which Kenneth Arnold's flying discs, and the major interpretation of them as alien craft, need to be placed. Some of the UFO references used in this brief review are classics, but others will be less familiar. Each is given the context in which it was interpreted.

Records of curious aerial phenomena have come down to us from the centuries BC, but I start at the beginning of the current era. Some centuries are skipped due to space considerations.

FIRST CENTURY AD

Pliny the Elder, the Roman natural historian, recorded in his *Historia Naturalis* a number of unusual aerial light phenomena known of in his time. Amongst several examples he mentions:

> A light from the sky by night, the phenomenon usually called 'night suns', was seen in the consulship of Gaius Caecilius and Gnaeus Papirius and often on other occasions causing apparent daylight in the night. In the consulship of Lucius Valerius and Gaius Marius a burning shield scattering sparks ran across the sky at sunset from west to east.
> [*Context: Supernatural, omens.*]

AD 584, 585

Gregory of Tours in his *Historia Francorum* records that in the year 584 'there appeared in the sky brilliant rays of light which seemed to cross and collide with one another' and in September 585 'rays or domes . . . which seemed to race across the sky' were observed by the French.
[*Context: Supernatural, miraculous signs.*]

AD 793

The *Anglo Saxon Chronicle* for this year reported:

> In this year terrible portents appeared in Northumbria, and miserably afflicted the inhabitants; these were exceptional flashes of lightning, and fiery dragons were seen flying in the air.
> [*Context: Supernatural, portents; zoomorphic, 'dragons'.*]

AD 989

Three brilliantly luminous round objects were seen on 3 August over Japan. They were seen to join together.[12]
[*Context: Supernatural, omen.*]

AD 1113

In this year a group of clergy from Laon travelled through the Wessex counties of southwest England, carrying supposed relics of the Virgin and performing miraculous healings. One Sunday, they encountered 'a dragon' at Christchurch which had come out of the sea 'breathing fire out of its nostrils':

> It was incredibly long and had five heads, from which it breathed sulphurous flames; it was flying around from place to place, and setting fire to houses one by one . . . .
>
> When the Dean had seen his house and church on fire, he had hastily collected his clothing and furniture and strapped them onto a ship which was beached in the harbour nearby. Then he had the ship launched and hoped that it would be safe from the fire. The dragon was nearby and . . . found the ship and flew over it, and burnt all that was on board.[13]
>
> [*Context: Religious, miraculous act of God; zoomorphic, 'dragon'.*]

AD 1170

The chronicle of Ralph Niger has an entry for 9 March, referring to a place called St Osyth, presumably that in Essex near Colchester (the scene of a major earthquake in the nineteenth century):

> . . . a wonderfully large dragon was seen, borne up from the earth through the air. The air was kindled into fire by its motion and burnt a house, reducing it and its outbuildings to ashes.
>
> [*Context: Supernatural, a wonder; zoomorphic, 'dragon'.*]

AD 1606

In May, fireballs were regularly reported over Kyoto, Japan. One appeared as a spinning ball of fire like a red wheel; it hovered near the Nijo Castle and had many witnesses.
[*Context: supernatural, portent.*]

MID-1700s

In 1756 traveller and geologist Alexander Catcott visited a copper mine near Bristol. The owner of the mine, a man called Pope, showed Catcott buckles made of his copper. He said he hoped to discover a large copper vein shortly because he had seen 'a large ball of fire, as big as a man's head' break out of the earth and shoot up into the sky. Catcott noted excavation taking place on the summit of the hill where Pope had seen his light emerge from the ground.[14]

It was widely believed up to this century that ores could be detected by the appearance of luminous 'vapours' above mineral veins. This is recorded by William Pryce in his *Mineralogia Cornubriensis* of 1778:

> Another way of finding veins, which we have heard from those whose veracity

> we were unwilling to question, is by igneous appearances, or fiery coruscations. The Tinners generally compare these effluvia to blazing stars, or other whimsical likenesses, as their fears or hopes suggest; and search, with uncommon eagerness, the ground which these jack o' lanthorns have appeared over and pointed out. We have heard but little of these phenomena for many years; whether it be, that the present age is less credulous than the foregoing; or that the ground being more perforated by innumerable new pits sunk every year, some of which by the Stannary laws are prohibited from being filled up, has given these vapours a more gradual vent; it is not necessary to enquire, as the fact itself is not generally believed.

This account not only records the belief in ground-associated light phenomena by early miners, but also demonstrates the 'Age of Reason' that came in over the seventeenth and eighteenth centuries, in which former observations tended to be dismissed as superstition, and everything had to be explained by logical science. In the following case we see how the terminology of the new age of reason was, in fact, no more logical than earlier concepts when faced with certain types of phenomena.
[*Context of first account: Exhalation of luminous discharges from minerals in ground. Context of second account: Superstition*.]

AD 1783

An extremely well-attested phenomenon occurred on 18 August. It was observed from the terrace at Windsor Castle by Thomas Sandby, Tiberius Cavallo, Dr James Lind, and Dr Lockman – all eminent men of the day – along with other witnesses. Cavallo gave an account of the sighting in the Royal Society's *Philosophical Transactions* the following year:

> To the northeast of the terrace, in a clear sky and in warm weather, I suddenly saw appear an oblong cloud moving more or less parallel to the horizon. Under this cloud could be seen a luminous object which soon became spherical, brilliantly lit, which came to a halt. It was then about 9.45 p.m. This strange sphere seemed at first to be pale blue in colour, but its luminosity increased and soon it set off again towards the east. Then the object changed direction and moved parallel to the horizon before disappearing to the southeast. I watched it for half a minute, and the light it gave out was prodigious; it lit up everything on the ground. Before it vanished it changed its shape, became oblong, and at the same time as a sort of trail appeared, it seemed to separate into two small bodies. Scarcely two minutes later the sound of an explosion was heard.

Not only were the witnesses of a high calibre, one of them, Thomas

Sandby, happened to be a famous watercolourist – a founder member of the Royal Academy. The event at Windsor Castle was duly painted. Sandby also produced aquatints of the event in association with his brother Paul, who was a topographical draughtsman employed by the Crown, and worked for the Ordnance Survey. A reproduction of one of the aquatints can be seen in Plate 1. It is the nearest we can come to a photograph of a UFO by a witness prior to the invention of the camera as we know it. [*Context: Age of Reason, meteor.*]

1896–7

There was an outbreak of 'airship' sightings over parts of America during this period. People saw lights like 'electric arc lamps' and 'balloon' and 'wheel' shapes, all of which were interpreted as sightings of a pioneer airship. Although some people came forward claiming to be inventors of the craft, these were never proved and were almost certainly hoaxes. There were definitely newspaper hoaxes connected with the wave of sightings. Claims of meetings with the (normal, human) airship occupants, and even rides in the craft, were made.

Airship technology was much in the news in Europe and America, but it is virtually certain that there was no airship available in the USA at the time capable of the performance necessary to explain the sightings. [*Context: Terrestrial technology, pioneer airship.*]

EARLY 1900s

The Irish writer Dermot Mac Manus recalled an incident he witnessed as a child of six or seven. It occurred near the Shanaghy crossroads between Kiltimagh and Bohola in County Mayo. It was twilight on the return from a family picnic, after a hot summer's day. He and his mother were being driven in a pony and trap. The boy Mac Manus was half-asleep when an exclamation from his mother and a wrench as the driver suddenly lashed the horse into greater speed jolted him back into wakefulness:

> It appears my mother saw this patch of ground covered with . . . dozens of little twinkling lights rather like fireflies. Most of them were about a foot or a little more off the ground and moved around with occasional bobbing up and down. They were all the same colour, a pale yellow, and sparkled in the prettiest way up and down over the mounds and in and out through the rushes. That they could be anything having to do with the fairies never occurred to her for a moment till the driver declared that it was the 'Shee' out

> and about, and 'The Lord protect us from them!' . . . But I saw them, too, and to this day I can still conjure up in my mind those little twinkling lights.[15]

Other people had also reported 'fairy lights' over the area at various times, Mac Manus later discovered.
[*Context: Supernatural, fairies/spirits.*]

1909

An outbreak of sightings of night-time lights in Britain took place between March and May of this year. Reports consisted mainly of lights, particularly 'searchlights' coming from unseen or sometimes dimly perceived flying objects. The British press and public primarily saw the lights as coming from German airships on spying missions.
[*Context: Terrestrial technology, foreign airships.*]

1930s

'Ghost planes' made their appearance over many locations at various periods in this decade, though most notably over Scandinavia.
[*Context: Terrestrial technology, foreign aircraft.*]

EARLY 1940s

Allied aircrew reported numerous contacts with a range of discoid and spherical objects, mainly over the European theatre of war, but also some over the Pacific. The objects were nicknamed 'foo fighters' and usually took the form of small orange balls of light, but sometimes small discs – some translucent – and occasional larger phenomena. They occurred singly and in formations. Although several aircrew reported apparent 'intelligent' behaviour by these discs and spheroids, they seemed insubstantial and never harmed aircraft. German pilots, too, witnessed these phenomena. In all cases, the main belief was that these lights and shapes were enemy radar decoys or secret weapons.
[*Context: Terrestrial technology, enemy craft.*]

1946

Almost a thousand reports were made of 'ghost rockets' in Swedish skies during this year. These were mainly light phenomena, but there were also numerous reports of cigar-shaped objects seen in broad daylight. Some 'rockets' were seen to crash into lakes with great explosions. Detailed searches made of a few of these lakes failed to reveal any form of wreckage. There was popular speculation that the rockets represented experiments by Soviet-captured German V-weapon expertise, but nothing checked out satisfactorily. The Swedes had had experience of re-assembling crashed V-rockets during the war, and were thus experts. This time, Swedish investigators could determine nothing. Some 'rockets' were seen coming from the Norwegian direction.[16]
[*Context: Terrestrial technology, foreign rockets.*]

Many sightings of aerial phenomena, the vast majority being *light* phenomena, have been reported since 1947. In the aftermath of the Cascades sighting, which, remember, occurred during a wave of sightings, numerous other discs or 'saucers' were seen, often by pilots or aircrew. The general context of witness and popular interpretation moved from secret terrestrial technology, to unknown technology, to extra-terrestrial technology: our own level of technological development made such an idea accessible to us. As the great psychologist Carl Jung wrote: 'The impossibility of finding an earthly base for the UFOs and of explaining their physical peculiarities soon led to the conjecture of an extra-terrestrial origin.' Almost every sighting that was reported was put in the context of extra-terrestrial craft by the majority of UFO researchers, the press, and the public.

Jung was well aware of the essentially phantom nature of human identifications of the UFO. He thought of UFOs as *projection carriers* in that they provided screens for the ideas, hopes or fears of the culture that perceived them in any particular age.[17] There was an actual, objective phenomenon, he put forward as one suggestion, which our physical eyes perceived but which was coloured by an internal projection belonging to the individual who witnessed the outer phenomenon, and thus belonging to the society, the culture, of which the individual was, in a sense, the product.

Today, many people feel the logic of science to be arid. It has exposed us to the vast impersonal domains of the outer space surrounding our tiny planet, a planet where our humanity seems to be threatened from so many directions. We seek celestial help, but God is dead – in a dynamic, cultural sense. Extra-terrestrials now occupy the throne of heaven. The extra-terrestrial projection onto UFOs is the scream of a lonely species.

In fact, as we can see from the above review, the extra-terrestrial hypothesis (ETH) is only one in a list of images that have been projected onto the

UFO, and does not possess any particular claim to be viewed as more likely to be 'true' than some of the others.

Arguments about the validity of the images imposed on the phenomenon are not new, however. The debate about whether lights in the sky were dragons or not, for example, started at least in the thirteenth century when Albertus Magnus wrote:

> Some people say they have seen dragons flying in the air and breathing sheets of fire. This I think impossible, unless they mean vapours of the sort described in the book *Of Meteors*, which are called dragons. These have been shown to burn in the air, to move and give off smoke, and sometimes (rolled into a ball) to fall into water and hiss as white-hot iron would; sometimes again the vapour rises in conditions of wind from the water, breaks out in the air, and burns plants or anything else that it touches. Because it ascends and descends in this way, and because of the cloud of smoke which drifts out on either side like wings, ignorant people take it to be an animal flying and breathing out fire.[18]

The argument simmered on over the years. In 1590 Thomas Hill referred to the skyborne, luminous dragon as 'a fume kindled' that took on the simulacrum of the mythical beast. In 1608 Edward Topskell wrote that the dragon should be understood as 'the Meteor called *Draco volans*' or fire-drake, and also in that century another authority considered the dragon seen in the skies as being 'a weaker kind of lightning'. In November 1792, a Scottish pastor noted that 'many of the country people observed very uncommon phenomena in the air (which they call dragons) of a red fiery colour, appearing in the north, and flying rapidly towards the east, from which they concluded, and their conjectures were right, a course of loud winds and boisterous weather would follow.' It seems these 'country people' knew their natural history well, whatever terminology they employed!

TO THE DRAGON'S LAIR

We have seen that some of the first glimmerings that there might be the possibility of terrestrial origins for unusual aerial lights came in the writings of Charles Fort. He noted that lights and earthquakes frequently occurred coincidentally, and could haunt the same geographical areas. The initial connection between geological factors and aerial phenomena considered specifically in the UFO frame of reference, however, seems to have been made by a Frenchman, Ferdinand Lagarde, in a 1968 article in *Flying Saucer Review*.[19] He made a study of a 1954 French UFO wave and noted that 37 per cent of low-level UFO sightings occurred on or close to geological faults, and that 80 per cent of the sighting locales were associated with faulting. To check if this was likely to be a chance effect, Lagarde tested the distribution of what he called communes in France against faulting and came up with a mere 3.6 per cent on faults, or 10.8 per cent if margins of 1½ miles (2.5 km) either side of faults were allowed. Further research showed a 40 per cent correlation between reported UFO incidence and fault lines. Lagarde commented: 'UFOs occur by preference on geological faults.' He suspected that more detailed geological information than that available to him during his research would probably enhance the correlation. Lagarde made the perceptive observation that geological faults 'are not merely the external aspect of an irregularity in the Earth's crust, but are also the scenes of delicate phenomena – piezo-electrical, or electrical, or magnetic, and at times perhaps of gravimetric variation or discontinuity'. In 1969 the Earth Mysteries writer John Michell endorsed Lagarde's view in his seminal work, *The View Over Atlantis*[20]: 'There is no doubt that . . . phantom lights are manifestations of electro-magnetic energy most commonly encountered in the neighbourhood of geological faults, during episodes of magnetic disturbance.' Michell went on to make a further point that ufologists became properly aware of only at a later date:

> the appearance of flying saucers is not entirely unrelated to the psychological condition of their observers. This is not, of course, meant as an implication that flying saucers have no physical reality or that those who see them are mad. . . . But it is a fact that people who see flying saucers often experience the sensations of a vision.

This is an aspect of ufology we will return to in the last chapter.

By 1970 veteran American ufologist John Keel was making connections ahead of his time.[21] His 1960s research had shown him that outbreaks of reported UFO events occurred in 'window' areas about 200 miles (322 km) across. These windows could also be part of larger geographical patterns, he felt. 'These are areas where UFOs appear repeatedly year after year,' he wrote. Keel had grasped the key factor: there was a *geographical* dimension to the phenomena. He further noted that 'many . . . reports are concentrated in areas where magnetic faults or deviations exist'. He also made another remarkably pertinent observation, as we shall see: 'UFOs seem to congregate above the highest available hills in these window areas. They become visible in these centers and then radiate outward . . . before disappearing again.' Keel referred to a paper by Dr Martin D. Altschuler entitled *Scientific Study of Unidentified Flying Objects* and contributed to Colorado University in which he suggested that the UFOs resulted from rock friction – the piezo-electric effect. At the time Keel felt such an explanation was 'as far-out an explanation as visitors from Mars', and proceeded to view the window areas as places where interaction with some other level of reality was taking place.

But Keel did make direct associations between earthquakes and light phenomena, and at the same time felt that mechanisms existed to allow such an association to take place at a global level, quite apart from the small, window context. He quoted the case of 'a brilliantly illuminated object' that flashed across the northeastern United States at around 8.15 p.m. on 25 April 1966. It illuminated the countryside as it passed overhead, was seen by a great many witnesses, and came 'right on the heels' of a nationwide UFO wave in America. Keel probed further. He noted some events in Tashkent, Russia, around 5.23 a.m., 26 April, 1966. Galina Lazarenko, a Soviet scientist, was awakened by a flash of light so bright it lit up her room enabling her to see all the objects in it. The courtyard outside was similarly brilliantly illuminated. At the same moment, engineer Alexei Melnichuk heard a loud rumble followed by a bright flash while walking down a street. 'I seemed to be bathed in white light,' he recalled. Then the great Tashkent earthquake struck. It killed 10 people and made hundreds of thousands homeless. Many people who rushed out of their homes in panic reported seeing balloon-like glowing spheres in the air. Thinking globally, Keel realised that the Russian earthquake *was occurring at the precise time the fireball was arcing through American skies*. Furthermore, Keel discovered that Tashkent is at the same latitude as the northeastern United States. Unless this was a meaningless coincidence to end all coincidences, the likelihood is that under optimum circumstances, stimuli can occur within the planet resulting in simultaneous geophysical effects on a grand scale in different parts of the world.

Between 1972 and 1976 Andrew York and I made a part-time, multi-disciplinary study of the central English county of Leicestershire. We carried out extensive field and archive research, noting records covering over four centuries of unusual meteorology, seismicity, the locations of ancient sites, traditional gatherings, reported supernatural and ufological incidents, and related these to one another as well as to the county's geology to see what patterns, if any, might emerge. One that came up (Figure 2) indicated that the faulting in the county, the areas where unusual seismicity and meteorology had been reported over the centuries, and the high incidence zones of reported UFO activity over a 25-year period, matched to a surprising extent. Although population distribution in the county confused this pattern to some extent, it certainly did not invalidate it. Leicestershire does not have a particularly exciting record of UFO activity, but we were able to perceive an underlying pattern indicating that those events which were reported did largely seem to relate to areas of faulting and certain kinds of exposed rocks over a period of years. One focus of UFO reports occurred round a curious hill called Croft Hill, some miles southwest of Leicester, which was a traditional gathering place down the centuries, for both religious and secular purposes. There is some evidence that it might have been the 'mesomphalos' of England and Wales, the central sacred hill, in Celtic times.[22,23] It is an outcrop of granite and syenite and has two minor faults associated with it. A quarry is situated next to it. Another important aerial phenomena focus in the county is Charnwood Forest, an area of very ancient exposed rocks which has its own internal system of faults as well as faulting around its perimeter. It has been an epicentre for numerous recorded earthquakes and tremors down the years, including the 1957 event (Chapter 1).

With this research, I started to become convinced that there had to be some sort of correlation between UFOs, strange weather and geological factors. Across the Atlantic, in North America, this correlation was given its most sophisticated presentation up to that time in *Space-Time Transients and Unusual Events* (1977)[24] by Michael A. Persinger and Gyslaine F. Lafrenière. Both authors were from Laurentian University, Sudbury, Ontario, where Persinger is a professor of psychology and a research scientist.

The two Canadians made a study of UFOs and other anomalous, 'Fortean' events (named after Charles Fort), primarily in North America, and subjected them to various computer statistical analyses. Results indicated that the basic idea of 'window areas' was probably correct. For example, they tested unusual event clusters in the state of Illinois and found that there was a particular area that over the years exhibited a relatively larger incidence of unusual events than anywhere else in the state.

They considered many potential mechanisms that might be involved in the production of unusual events, particularly UFOs, ranging through stellar, planetary, solar and lunar influences on the Earth's geophysical systems, such as the geomagnetic field. The authors emphasised the vast energies contained within our globe: 'the existence of man upon a thin shell beneath which mammoth forces constantly operate, cannot be overemphasised.'

Such combined extra-terrestrial and terrestrial processes, Persinger and Lafrenière argued, would affect large areas of the globe, but might result in outbreaks of lights or unusual phenomena only in certain locations where geological stresses and other terrestrial factors were in a state of tension, ready to be triggered. (Tectonic stress waxes and wanes throughout the Earth's crust in many places every day, producing many small – sometimes imperceptible – tremors and only occasionally erupting as a major earthquake.) Fields of forces operating over very large geographical regions might focus in on just a few small areas due to particular seismic tensions, rock and mineral distributions and the like, creating localised conditions in which outbreaks of phenomena could take place that would be unlikely to occur if the background forces were evenly, widely distributed. They likened this to a stiletto heel making an indentation on a hard floor, whereas

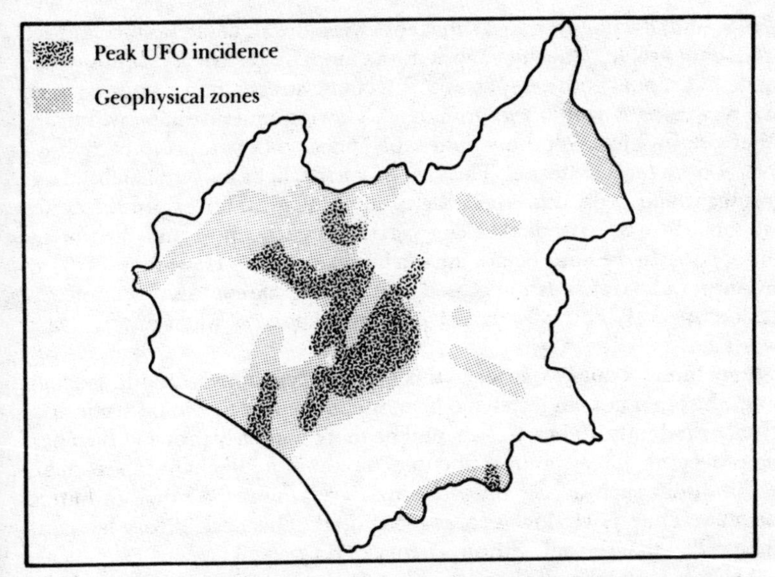

FIGURE 2 The county of Leicestershire, central England. Pale grey areas are regions of unusual meteorological or seismic activity as recorded over more than four centuries; dark areas are regions of highest reported UFO incidence.

the same force applied through a flat-soled shoe would leave no mark. Or, again, the energies in the atmosphere that produce a gentle breeze over a wide area can sometimes concentrate into the localised ferocity of a tornado.

The Canadian researchers also pointed out that extra-terrestrial forces, such as charged particles from the sun, acting through terrestrial mechanisms, might trigger phenomena only after a time lag. But, whatever the hinterland of forces involved, Persinger and Lafrenière felt that the evidence provided by their research pointed to the immediate cause of light phenomena being 'seismic-related sources'.

Seeking a model for a specific mechanism that might translate tectonic stress into light phenomena, the two researchers opted primarily for a modification of the piezo-electric effect (see Chapter 1) which had been proposed in 1970 by Finkelstein and Powell[25] amongst others for the production of earthquake lights (EQLs). Persinger and Lafrenière suggested that forces accumulating in a seismic area, perhaps over weeks or months, might produce a localised 'electric column'. Such a column would normally have a radius of between 10 and 100 feet (3–30 metres) but could, conceivably, have a radius of up to a mile under certain circumstances. The high electric field within such a column might cause a number of Fortean effects, and, if values reached sufficient levels, the air could become ionised producing visible, glowing shapes in the air. The electrical column would move in keeping with the passage of tectonic stress along a fault or other line of weakness within the ground. A glowing ionised shape within the otherwise invisible column would thus appear to a witness to be flying a course over the landscape. The size of such a light and its height above ground would depend on the scale of the geological forces producing the column. (Strong earthquakes can certainly reach up to and disrupt the ionosphere. In the hour before the earthquake at Hilo, Hawaii, in 1973, for instance, radio transmission ceased due to the apparent 'disappearance' of the ionosphere!) This effect could work over bodies of water just as well as over land.

Some areas would experience this only on cycles measured in geologic time, thus to a human timescale phenomena appearing would seem to do so only randomly. Other areas would be more regularly prone to such flexing of tectonic forces, and so phenomena are more likely to repeat themselves within such locales on cycles that are within the range of human memory. Thus a 'window area' is recognised. This basic theory has come to be called the Tectonic Strain Theory (TST).

The two researchers pointed out that earthquakes themselves would not be necessary to produce such lights; indeed, a quake would release the stress being built up within rock and mineral bodies within the crust. A quake only occurs when rocks give way under the strain, otherwise the stress builds up and then relaxes, the massive pressures involved not actually triggering deformation of the Earth's crust. Persinger and other proponents of the TST have suggested, therefore, that lights might be associated with very small quakes, where only limited rock fractures have occurred. Such seismicity might not register sensibly to people over such locations of tectonic stress, and may, indeed, not even be recorded by the available network of instrumental monitoring.

The theory also allows for effects on the human witness of light phenomena. The electric column could affect the brain function of a witness standing within it while observing a lightform. This would allow the possibility of explaining 'exotic' UFO encounters where humanoids, abductions and short periods of amnesia are allegedly reported. Further, energy fields surrounding light phenomena may cause burning and radiation-like effects on witnesses, as well as on the immediately local physical environment. Such effects are reported from time to time in the UFO literature.

In 1980, the Swede Ragnar Forshufvud, writing in *Pursuit*,[26] also suggested that UFOs were a phenomenon akin to ball lightning but produced by the Earth. He suggested that UFOs are 'cells of oscillating electromagnetic energy' generated from the action of *earth currents* under particular geological conditions. He seems to have independently arrived at a very similar conclusion to Helmut Tributsch who had already been working on similar ideas. Earth currents, created by the piezo-electrical release of energy, pass through the Earth's crust by dependency on ions and other charge-carriers in groundwater squeezing through porous rocks or interstices and channels between rock masses. It is thought that earth currents (or geoelectricity) are linked in some way with currents flowing in the ionosphere.

In the days when low-frequency telegraphy was used, earth currents could interfere with connections. Today, they are part of the range of earthquake precursors monitored by geologists, because changes in geoelectricity in an area can occur in the build-up to a quake, along with alterations in the local geomagnetic field. Forshufvud suggested that 'deep fractures open in the path of the current, acting as electrical circuit breakers, forcing the current to change direction. In the fractures, flashover is likely to take place.' Highly compressed gas, such as methane, can fill fractures, and voltage passing through such gas-filled space 'may rise to comparatively high levels before discharge takes place'. He estimated that enough energy could be generated by such means to produce lightballs, in the manner that atmospheric charge building in a thunderstorm can allow ball lightning to be produced.

Forshufvud felt there were three basic questions that had to be answered in the affirmative if the theory was to be considered as a possibility:

1 Are UFOs often seen along fault lines?

2 Are UFOs often seen in connection with earthquakes?

3 Is there a correlation between geomagnetic and UFO observations?

The first question will be positively answered by this present book, the second can certainly be answered 'yes' as we have seen in Chapter 1, and Forshufvud was able to provide some evidence for the third. He took UFO data that had been specially 'screened' and selected by the arch-ufologist, the late J. Allen Hynek, who conducted research into and evaluation of UFO reports for decades, and who coined the now famous term 'close encounters'. These Forshufvud ranged against sunspot cycle data, and, as can be seen from Figure 3, he discovered that there were noticeable rises in reported UFO activity close to the times when most sunspots occurred (*sunspot maxima* – a roughly 11-year cycle) in the period the data covered. The French researcher Dr Claude Poher had already found correlations between increased reported UFO activity and rises in geomagnetism by this time, as well.

The tentative findings from the Leicestershire work intrigued me after the study ceased in 1976. When the opportunity of writing a book on the subject was presented to me in 1979 I jumped at the chance. With contributions from trained geochemist Paul McCartney, *Earth Lights* was pub-

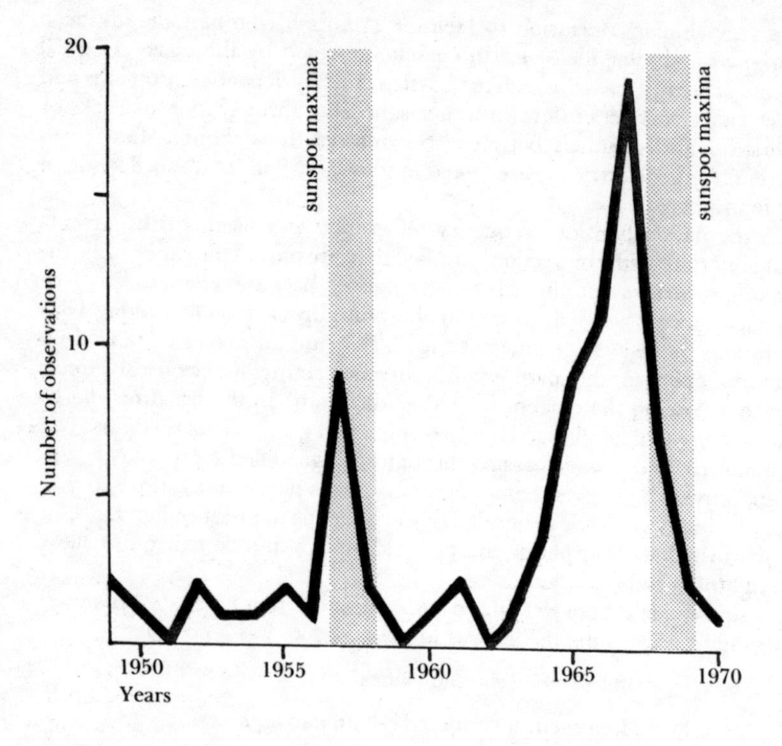

FIGURE 3 Swedish researcher R. Forshufvud compared 20 years of UFO reports with sunspot activity. It can be seen that peaks in both sets of data closely relate. (*After R. Forshufvud.*)

lished in 1982.[27] It was only in the research for this that I became aware of the work of Persinger and others in North America, though I was already aware of the work of Lagarde and Keel. It seemed to me that we needed to close in on much smaller areas than had been attempted at that time in the North American studies to see if there really was an intimate connection between reported UFOs and faulting: while faulting was by no means the only geophysical aspect to take into account, it was fair to expect that faulting would frequently be a prominent factor in areas associated with light phenomena if the basic TST was correct. It was one important aspect of the pattern to check. Patterns that look convincing when brushed in with broad strokes, can, when examined at close quarters, break down. We were to find that in this case the opposite was to occur – it was to get stronger the closer we looked. We were in a fortunate position: Britain is a relatively small area, yet possesses 'amazingly enough, strata of every geologic period from the pre-Cambrian to the Quaternary' as one Japanese book puts it. Added to this, Britain is very well mapped both geographically and geologically (though there are still annoying gaps in the readily-accessible large-scale geological record), and it has active UFO research groups so a reasonable collection of reported UFO data takes place.

Even with these advantages, however, it proved surprisingly difficult to get an area that had a good set of UFO reports from which geographical locations could be accurately determined, and which also had sufficiently detailed geological maps to at least 1:50 000 (2 centimetres to 1 km) scale.

But our first 'test' was both obvious and easy – Warminster in Wiltshire. This small market town came into ufological prominence in the mid-1960s when a 'Thing' started to be seen in its local skies. People from all walks of life saw lights and cigar-shapes in the area. A local journalist, Arthur Shuttlewood, began to write colourful books on the matter, and UFO events reported at Warminster were regularly described in the UFO specialist journals (Warminster even had its own for a while), and the general press, too. Before long, UFO enthusiasts from all over Britain – and elsewhere – began to descend on the area to carry out 'skywatches' on nearby hills: Warminster had become one of the country's most celebrated 'ufocals'. Even the town's name derived from 'worm' which meant serpent or dragon in old English.

One of the key locations for UFO manifestation was Cley Hill, a mile or two to the west of Warminster. It was a chief 'window' according to Shuttlewood when I asked him at the time. A curious natural hill, it was also artificially earthworked in prehistoric times.

For our research, we contacted Barry Gooding, a local ufologist who had kept a careful eye on the UFO reports over the years. He felt, like most serious ufologists, that the great majority of the supposed sightings were of aircraft lights, satellites and other misidentified mundane features. Furthermore, there had been a good deal of media 'hype' surrounding the Warminster Thing. Nevertheless, he was quite satisfied that there was *something* odd going on. He felt a certain percentage of the reports did refer to a real, but unexplained, phenomenon – an orange lightform that was seen to gambol around the local Warminster landscape. It came to be nicknamed 'The Amber Gambler' by the ufologists in the area. Gooding also had accounts from people who had seen lightforms in ball and ellipsoid forms apparently *entering the slopes of Cley Hill*. Other reports talked of lightbeams emerging from the ground, forming into balls of light and flying off.

Warminster looked an improbable candidate for the earth lights theory because it was situated in chalk country (as demonstrated by the nearby Westbury White Horse chalk hill figure, for example) and such geology is usually tectonically stable. So there were unlikely to be many surface faults around. When the 1:50 000 scale geological map was consulted this was the

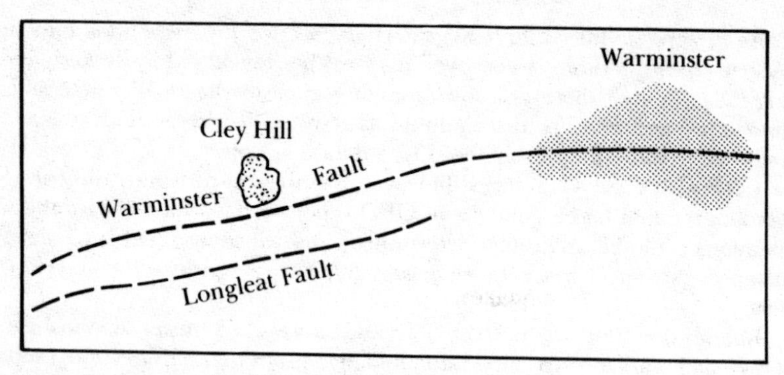

FIGURE 4 Faulting near Cley Hill and Warminster, scenes of intense reported UFO activity in the 1960s. There are no other recorded faults in the region.

case – there were only two recorded surface faults in the whole region of countryside around Warminster. But those two were right at Warminster itself! (Figure 4.) One fault goes immediately alongside Cley Hill and on through the town, while a shorter fault runs parallel to the first, close to Cley Hill. We had scored an unlikely bullseye.

We looked at other locations where we had a workable match of usable UFO data and suitable geological information available. We will revisit a few of these later in the present book, as new information has become available.

Earth Lights also tried to analyse Hynek's classification of reported UFO types from the viewpoint of the earth lights theory. Hynek had broken UFOs down to 'nocturnal lights'; 'daylight discs'; 'close encounters of the first kind' (CEI) – close observation of UFO but with no material interaction between UFO and witness; CEII – where the UFO has some direct effects on witness, the immediate environment, or both; and CEIII – those in which the presence of animate figures is reported. Cases were used in *Earth Lights* to illustrate a slightly modified version of this classification. (This was criticised by some when the book came out, as the cases were taken at face value as reported, and were not investigated individually by me. This was beside the point, of course, because they merely illustrated the *types* recognised by Hynek – and there were in any case detailed studies elsewhere in the book.) Essentially, it was possible to see that the majority of UFO reports, in fact just about all of them in one way or another, were of what John Keel called 'soft' objects: at some point in their manifestation they did not behave like hard, solid machines or forms.

Earth Lights also attempted to initiate the opening up of a number of cross-connections between UFO phenomena and other factors, and to give the first comprehensive account of earth lights theory up to that date.

In the same year, 1982, another book relevant to the area appeared –

When the Snakes Awake[28] by Helmut Tributsch, a German biophysicist. Tributsch was fascinated by accounts of animal sensitivity to earthquakes – in many cases it seems dogs, cats, chickens and other creatures could sense the approach of a quake. What could the mechanism be? In the course of the 1970s, Tributsch looked at many earthquake effects such as light phenomena, how magnets were sometimes reported to lose their magnetism prior to or during an earthquake (regaining it afterwards), strange fogs, sky colourings, sounds and odd weather effects associated with such tectonic events. His trail eventually led him to consider the electromagnetic changes in the ground preceding quakes.

Tributsch demonstrated the extraordinary sensitivity of animals to changes in environmental electrical fields. He came to the conclusion that prior to an earthquake the ground emitted electrostatic charges. Such charges, Tributsch reasoned, become attached to aerosol components suspended in the air. These were breathed in by animals which then became charged up themselves. This ultimately led to their disturbed behaviour, perhaps because of effects of the inhaled charged particles on serotonin, a nerve hormone in the lower middle brain. But how was the electrostatic emission generated in the ground in the first place? The obvious candidate was piezo-electricity. Like Persinger, Tributsch looked at the various theories regarding this, especially the ideas of Finkelstein and Powell. There was no doubt that piezo-electricity could supply the energy to produce electrostatic charge in the air, indeed to produce glow effects also, but the problem was that the ground's resistance is sufficiently low to equalise charge built up within it, thus preventing any detectable energy being emitted into the atmosphere. Tributsch opted for a mechanism called *electrochemical glow discharge*. Stated very simply, this could be achieved by earth currents, flitting through the ground along the lines of least resistance, passing through thin films of water in tiny rock fractures. Through various electrochemical reactions the expulsion of considerable amounts of gas into the air (bubbling through groundwater or out of fissures in the ground) accompanied by pungent chlorine or sulphurous smells could be caused. Such smells have been noted by witnesses involved with earthquakes, ball lightning and UFOs. Explosions and fires might also be caused by this process. Further, the passage of current from a comparatively conductive solid across a thin film of air in hairline cracks and into water beyond would cause a glow discharge, a poorly understood effect. Reactions caused by glow discharge resemble those caused by ionising cosmic radiation, X-rays and low-energy alpha radiation.

The ideas proposed by Tributsch were resisted at first by the scientific community, and he had a number of papers turned down by scientific journals. Eventually, *Nature* did publish his work (in 1978) and this led to communication from Stuart A. Hoenig of the University of Arizona who had noted that rocks subjected to high pressures in his laboratory did emit electrically charged particles before bursting. Hoenig considered this the first experimental proof of the existence of electrostatic earthquake precursors.

The year 1983 saw the publication of Jenny Randles' book *The Pennine UFO Mystery*[29] which also noted the existence of significant geological faulting in the areas of reported major UFO events. In the same year an article by Paul McCartney, Dr Don Robins and myself appeared in *New Scientist*[30] which developed a few aspects initiated in *Earth Lights*.

Considerable further research, on both sides of the Atlantic, has now taken place, and more thought has been expended on the earth lights idea. Aspects of these developments form the material for the rest of this book.

Electricity, magnetic changes, the production of gases, the emission of radio waves, sounds and other energy effects are clearly associated with tectonic pressures. It is a promising sea of forces, in many ways uncharted by science, from which EQLS certainly, and earth lights probably, emerge. Furthermore, such seismic conditions may create electromagnetic circumstances similar to those occasionally engendered in electrical storms – a number of researchers such as Tributsch have made this point. This could mean that ball lightning, EQLS and earth lights share some basic relationship, even if they are individual phenomena with certain special characteristics, like members of a family. But this still leaves us facing light phenomena whose precise nature still eludes us: while the matrix of conditions they appear out of may be within the bounds of known science, we are looking at an end-product that is challenging, and must ultimately extend branches of our understanding beyond their current reach. The mighty forces of our planet, probably prodded and persuaded by even mightier extra-terrestrial influences, seem to combine in certain ways under certain conditions to produce completely unfamiliar forms of energy or tenuous matter. Unravelling the nature of that energy might lead us into revelations that will shake current conceptions of geophysics and consciousness to their foundations.

The ancient Greeks thought earthquakes were caused by a mysterious vapour, *pneuma*, issuing from the ground; Japanese miners noted a 'strange air' they called *chiki* coming out of the bowels of the Earth prior to quakes, and modern scientists tell us the earth releases electrical charges during tectonic stress. Perhaps the medieval observers, in believing that the fiery dragons came from deep caverns, were more perceptive than our times give them credit for.

THE CASCADES REVISITED

It is now time to return to Kenneth Arnold and his 1947 experience, and look at it with different eyes. I feel that it tells us more about a light phenomenon originating here on Earth than ever it does about extra-terrestrials.

First, what initially caught Arnold's attention? A 'tremendous flash' which 'appeared in the sky', he tells us in his 1977 account, 'a tremendously bright flash' in 1952 and 'a bright flash . . . it startled me' in 1947. Then he saw the light again 'almost like an arc light', and traced it to a formation of objects many miles to the north of Mount Rainier. Arnold's frame of reference at the time was that the light *could only* have come from the bright metal surfaces of an aircraft. Hynek was rightly puzzled how reflections from objects so far away could cause such a brilliant light. This is even more the case, because Hynek was labouring under the delusion that the distance at which Arnold first saw the objects was about 25 miles (40 km), but in 1952 Arnold made it clear that they were 'at a distance of over a hundred miles' when he initially spotted them. Whatever the brilliant flash was that lit the skies around Kenneth Arnold, it was not reflections from the flying discs.

As Arnold looked at the row of 'very bright objects' (1952 account), they 'fluttered' and tipped 'their wings' emitting 'those very bright blue-white flashes from their surfaces'. From wing surfaces? Arnold's frame of reference shows here – he speaks of the objects' 'wings', yet his drawings and closer descriptions show they were discoid. *They did not have wings*. He noted that they would 'give off a flash' before they gained altitude or deviated in their flight path. It seems Arnold may have had some later doubts about the nature of the light coming from these already very bright objects: in his 1947 account, he simply refers to the light flashes as reflections; in 1952 he said 'at the time' he assumed the lights to be reflections, and in 1977 said 'I *at first* assumed [the light flashes] came from their surfaces' (my emphasis). This can only mean that he came to think more deeply about this interpretation. He recalled, also in 1977, that the flashes of light 'would pulsate'.

We have to strip Arnold's account of its unsupported terminology of aircraft, wings, reflections, 'echelons', and so on to get outside his 1947 frame of reference. What did he *actually* see? A chain of spheroid or discoid *light-forms* surely – there is actually nothing else in Arnold's description than that, if one takes away his initial and overriding 'assumption' – his own word – that they were 'aircraft', and all the other assumptions and terminology which followed from that. The lights 'flipped and flashed' (1952 account) 'very close' to the contours of that part of the Cascade range. Remember the 'straight row of round masses of light' seen during the Idu peninsular quake of 1930 (Chapter 1). Also, as the objects flew past Goat Ridge, one of them turned causing Arnold to think that the objects looked 'rather like tadpoles'. Recall the lines of 'tadpole-shaped lights' seen in the skies above Leicestershire and surrounding counties at the time of the 1957 Charnwood earthquake (Chapter 1). This tadpole shape with regard to strange lights also recurs in unexpected circumstances: medicine men in one group of North American Indians in the Great Lakes area, for example, used to conjure up spirits at certain places that appeared as lights 'much larger than a man's fist' which had 'long tapered tails' according to descriptions given to missionaries, while in Japan there were said to be *shito dama* – 'fireball spirits'. These came in two basic forms: 'square fronted' and 'roundish tadpole shape'.

Arnold's discs were not terrestrial aircraft. And the evidence says nothing at all about extra-terrestrial craft. What it actually *does* tell us is that a brilliant aerial flash of light occurred, centred on an area where Arnold saw nine bright lights. These travelled in an undulating fashion between (in all) three peaks of the Cascade mountains, hugging the terrain, frequently giving off brilliant blue-white flashes, usually preceding a change in motion. As Arnold obtained a closer view of them, the lights showed themselves to be generally discoid in shape, though with variations which may have been due to shape-changing characteristics, or difficulty of definition at long distances on Arnold's part. The lights varied in intensity, and maintained a rough formation.

But there is more to support the earth lights interpretation of Arnold's discs than simply his accounts. The *location* where the incident took place also speaks volumes. The Cascade Mountains are located directly on a tectonic plate margin. Further south, this margin is marked by the San Andreas Fault. Tectonic plates are semi-rigid slabs of the Earth's crust floating on oceans of magma (Figure 5). Where these plates bump and jar against one another, or one edge slips and grinds beneath another, mountain chains are built, volcanoes erupt, earthquakes occur. Most of the plate margins forming the periphery of the Pacific basin are *destructive plate margins*, causing the infamous 'ring of fire', so-called because of the number of active volcanoes on it. The Cascades are on such a destructive margin, between the North American and the Pacific plates.

And we know the Cascades are tectonically active: this was graphically displayed on 18 May 1980, when Mount St Helens (only about 60 miles – 97 km – from Mount Rainier) erupted, blowing away the north face of the mountain and causing severe damage up to a 16-mile (26-km) radius around it, with the loss of 61 lives. Persinger has stated that he and co-researcher Lynn Henry were 'getting reports of luminous displays and other Fortean (anomalous) phenomena from Mount St Helens *before* the eruption'.[31] It had started rumbling earlier, in March 1980, when airborne geologists saw 'eerie blue lightning' arcing between the main crater on the volcano and another which had appeared since the beginnings of the disturbances.[32] It was earthquakes that triggered Mount St Helens's big blow-

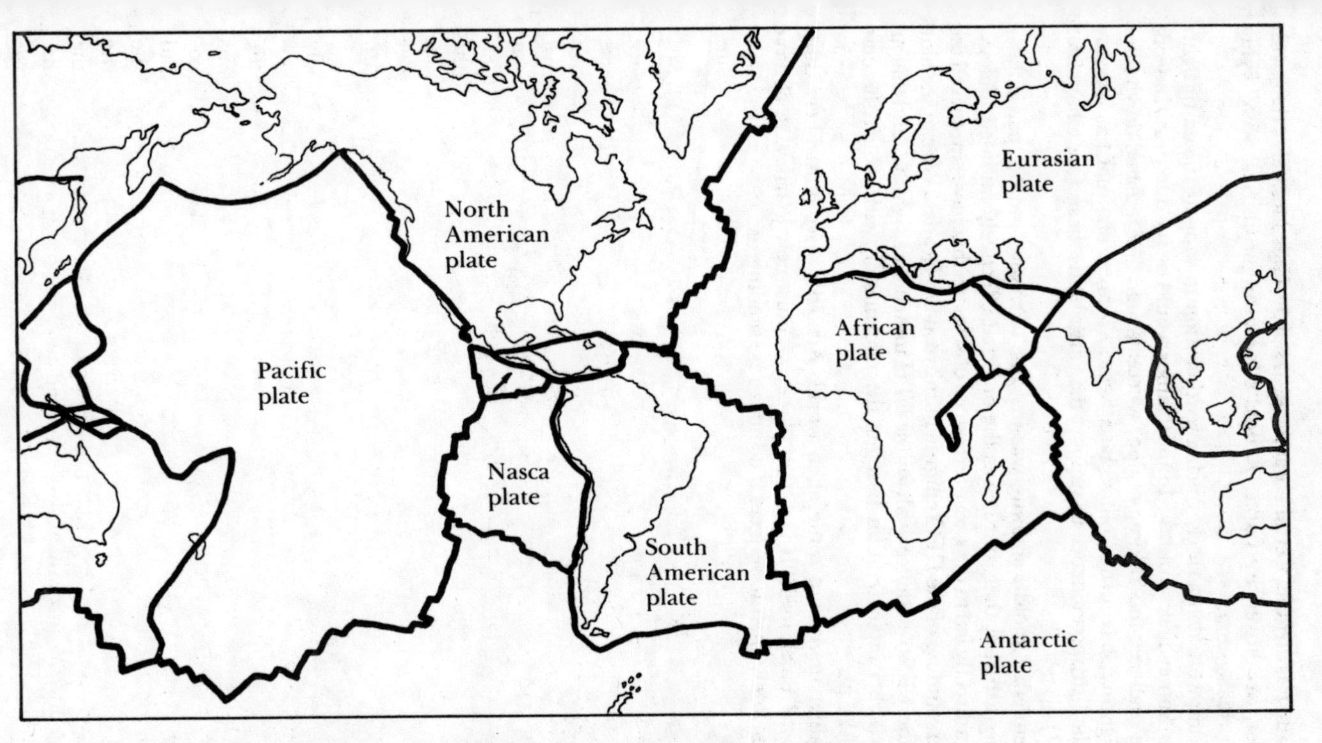

FIGURE 5 Simplified diagram of the pattern of tectonic plates making up the world's crust. Bold lines represent plate margins, zones of increased seismicity.

out, and it is virtually certain that the peaks of the Cascades, all of them volcanoes, are linked by faults. Arnold's 'very bright objects' were flying along this system.

If one were deliberately to choose a location to support the earth lights thesis, somewhere around Mount Rainier would be hard to beat. Indeed, as we shall see, the Yakima region, where Arnold first touched down after his sighting, has turned out to be a 'window' area, one that has recently supplied strong evidence to support the earth lights thesis (see Chapters 5 and 7).

Kenneth Arnold's sighting was just one of many, and not particularly distinguished at that, but it was the one that effectively initiated ufology, and indirectly led to the extra-terrestrial obsessions that have smothered the understanding of the UFO enigma in subsequent decades. We can see here that there is no evidence at all for the ETH in Arnold's sighting, and that an earth lights interpretation best fits the information with which the case provides us.

Rather than from some alien world, Arnold's discs most likely came from the Earth itself. It is to that planet we must now journey, to observe more closely the messengers it seems to be sending us.

PART 2
LIGHTS IN THE LANDSCAPE

3 BRITISH LIGHT PHENOMENA

The chapters in this section of the book will take the regional characteristic of earth lights as a theme. As Allen Hynek wrote in *The UFO Experience* (1972) regarding some classes of UFO: 'Localization is . . . a salient characteristic. . . . There is little tendency for the UFO to "cruise about the country" except locally.'

The following selection does not pretend to be complete – such a undertaking would in any event be impossible. There must be many thousands of earth light zones around the world. No one has the full record; indeed, hardly any attempt has been made to compile one at all. The geography of light phenomena has slipped through the fingers of scientists and ufologists alike.

We start our partial survey in the British Isles. (See Figure 6.)

SCOTLAND

Scotland's most famous anomaly is the Loch Ness monster, butt of jokes and a target for journalists in the 'silly season'. The loch lies along the major Great Glen Fault. Light phenomena have not been spectacular over Loch Ness, but they have quietly haunted the place over the years. Two lights were seen on 19 September 1848, and the following year another two inexplicable lights were seen over Inverness at the northeast end of the loch. 'Ghost ships' were seen from time to time on the water during the early decades of this century. The dedicated monster-watcher, Tim Dinsdale, saw a light shaped 'like a Chinese lantern' in the sky near the loch in September 1969 and, in another instance, a light emerge from a hillside. Lights were also seen on different occasions in 1973 and in 1978.[1]

But other lochs can lay claim to occurrences of light phenomena too – some of them on a more distinct basis. Lights were reported skimming along the waters of Loch Tay (northwest of Dundee) in the 1930s; tongues of flame were occasionally observed at Taagen close to Loch Maree (east of Gairloch

FIGURE 6 A geography of light phenomena – earth light zones mentioned in this chapter. Black stars represent localised areas; white stars regional areas of earth lights activity.

on the west coast), and at Upper Loch Torridon (by Shieldaig on the west coast) a light was frequently reported in the first half of this century emerging from a particular bay on the south side of the loch near Shieldaigh known as Ob Mheallaidh. Lights were also frequently seen travelling from the hills above Gruinard Bay (on the west coast) near the village of Coast to an islet on the other side of the bay called Fraoch Eilean Mor. When a light reached this rock, it would suddenly disappear. It was described by one witness as resembling a ball of light 'without rays'. Another witness experienced 'a strange noise' in his ears when the light passed close to him on one occasion.[2] (Such descriptions should suggest that they belong in the annals of geophysics rather than those of folklore.) In 1885, a light appeared frequently on Loch Rannoch during the months of October and November and was witnessed by many locals, who feared that it augured some dreadful happening. Two men rowed out to investigate the light, but lost sight of it as they approached, though it was still quite visible from the shore. In the 1950s writer Alasdair Alpin MacGregor told of a ball of light that was frequently seen skimming over a particular stretch of Loch Rannoch. Numerous other lochs seem to have boasted light phenomena, too, though records of them are hazy. And it still goes on: 'eerily white lights' were seen at Campbeltown at the head of Campbeltown Loch on the Mull of Kintyre as recently as March 1987, for example.

Lights are also seen from time to time in areas immediately adjacent to the myriad lochs of the Scottish west coast – particularly the Oban–Mull area. The 'very large ball' of orange light trailing 'a tail of flame behind' seen moving in from the direction of the sea near the village of Benderloch on 18 March 1988 by two locals, for example, is typical of accounts in the region.

The Scottish site most distinctly associated with earth lights, however, is Loch Leven. As the account given in Chapter 1 suggests, this sea loch, about 10 miles (16 km) south of Fort William, seems to have become quite famous around the turn of the century for the lights that were seen to race along its northern shore. A correspondent to *Folklore* in 1901 wrote that 'the lights abound' at the loch:

> There they are seen (by educated Lowlanders, too) on the Isle of St Mun, an old place of burial, and on the opposite side of the loch [from Glencoe], on the road to Callert. When I was at Carnoch House last year, opposite Invercoe, an English friend of mine observed the light closely, and about 10.30 p.m. in late August, the Ballachulish villagers turned out to stare and wonder. The lights moved rapidly down the road to Callert, then climbed the hillside, then went down to the shore of the loch. My friend could form no theory to account for their nature and movements, which are rapid. The country people have various hypotheses, all supernormal. No doubt there is a natural explanation, but so far, conjecture has been baffled.

At my request, Chris Garner of Edinburgh checked the Loch Leven area to see if strange lights are still being seen. It seems there is now no great incidence of 'the Callert Light', but lights are still observed sporadically. In the summer of 1982, Mrs Scott (pseudonym) was driving with her four children along the north shore of the Loch towards Onich at about 5 p.m. Shortly after passing Kinlochleven school, she saw a 'red, sun-like light' on the opposite side of the loch, about half way up the side of the mountain known as Ben Garbh. The light moved up and down, and then came towards the woman and children. She and the children were terrified. Mrs Scott stopped the car. The light also stopped. As she drove off again, the light likewise began moving. This procedure happened a second time. Halfway along the loch, Mrs Scott pulled up under a tree for safety. The light gathered speed and moved off over the hills in a southerly direction, towards Oban. Total duration of the sighting was estimated at between eight and ten minutes. (Witness interviewed in June 1988.) Twelve years earlier – November 1970 – on the same road, a taxi driver driving through a stormy night to collect a client encountered two lights about two miles east of Callert. Although they looked too yellow to be car headlights, the taxi driver assumed that was what they must be. They approached his vehicle along the road and as the 'car' was about to pass the taxi, the lights simply extinguished themselves. The taxi driver could see no trace of a vehicle, and his 'hair stood on end'. The man later spoke to a local, elderly bus driver who told him he had had similar experiences. (Witness interviewed June 1988.) Garner also managed to collect reports of lights being seen during the last 20 years from other witnesses.

The Scottish Western Isles also have tales of strange lights. On the Isle of Lewis, for instance, a light was seen 'for generations' a couple of miles southwest of Carloway. It would follow a specific course of about 200 yards and vanish when it reached an old well.[3] This light was associated locally with a tradition telling of the ghost of a murdered pedlar. In Loch Carloway, close by, a strange light was seen on several occasions during the 1940s; it seemed to home in on a motor boat that was left at anchor in the fiord. Lights have been reported, too, in the Holm area of Lewis, near Stornoway, and at other points on the island. Mull has a rock called Carraig na Soluis, indicating it to be a stone associated with the appearance of lights. Other islands, like Luing in the Firth of Lorne southwest of Oban, or, particularly, Benbecula in the Outer Hebrides, have yielded lights reports from time to time.

On the eastern side of Scotland there are a few recorded locations where lights have been reported at certain times. The East Coast fishermen have seen lights on numerous occasions on the Moray Firth and the Aberdeenshire (Grampian) coast. 'Ghost lights are so frequently seen by them,' Alisdair MacGregor asserts, 'that they tend to be regarded as quite ordinary, though perhaps infelicitous occurrences.' A key area seems to be clifftops on Gamrie Bay near Gardenstown on the north Aberdeenshire coast. Further north along the eastern coast, at Latheronwheel, close to Latheron, south of Wick, a 'bright, though somewhat diffused' light was seen on several occasions on a high point near the coast by witnesses in different locations around the spot.

The Scots give the Gaelic name of *gealbhan* to this type of light. Such luminosities have never been properly investigated. This is a pity, because in the next chapter we will see the enormous store of information that can be uncovered when these 'ghost lights', dismissed to the domain of folklore, are stripped of local belief systems and treated seriously as actual phenomena – which they are.

NORTHERN IRELAND

A focus of terrain-related lights in the province seems to be the Lough Erne area, near Enniskillen, Fermanagh. Lights there came to general notice when the Earl of Erne wrote an account in the *Daily Mail* of 26 December 1912. He stated that 'This light has been seen at intervals several times within the last six or seven years' on Upper Lough Erne and went on to give the following description of the form taken by 'the' light:

> It is of a yellow colour, and in size and shape very much the same as a motor-car lamp. It travels at a considerable pace along the top of the water – sometimes against the wind, at other times with it. It lights up all objects within a certain radius and disappears as quickly as it appears. It is mostly seen on stormy and wet nights rather than on fine ones.

A little later, the Countess of Erne wrote of an encounter she had with a light on the lough on 'Easter eve' in 1910 at about 11.30 p.m.:

> I saw a light crossing the lake below the windows of Crom Castle. It was . . . quite round, and about two feet across. . . . Its colour was a deep yellow. Its peculiarity was that it threw no light behind, but in front there was a blaze; so much so that when it passed a small copse, on the borders of the lake it lit up the trees, showing each trunk clear and hard.[4]

Letters in the Press elicited numerous further accounts, indicating that lights had been seen from time to time on both the lower and upper loughs by numerous witnesses. One particularly interesting account of lights on Lower Lough Erne was given by a George Irvine who recalled an experience as far back as March 1849:

> . . . as we came down to Slavin church, coming from Belleek, we saw a couple of very large lights under Slavin rock on the lake . . . we had a good view of the lights [one of] which seemed to go out, the other shot across the lake like a shot from a gun; it came back quite slow, till it came within a few yards of the road and shot again under a place called Maho Snap. It came up quite close to the road, and threw a light.[5]

Irvine also commented that he had 'often heard old folks talk of the lights on Lough Erne from Maho Church to what we call Morrow's point near the Holm'.

The recurring light phenomenon on Upper Lough Erne has come to be known as 'The White Light of Crom'. Mary Rogers in her 1980s book *Prospect of Fermanagh* records that Major Henry Cavendish-Butler, formerly of Inish Rath, saw the light on two occasions, both times 'on the lake, behind trees, which were lit up to the topmost branches'. A recent owner of Inish Rath island, who bought it from the trustees of the Butler family, has also seen the light, and chased it in his motor launch – without success.[6] Rogers also notes the account of a former head gardener at Crom, who saw the light on the middle of the lake from the quay. He described it as 'a beautiful golden colour' and remarked that 'there was no reflection on the water, or rays from the light . . . the light appeared to be quite steady, and only a foot or so above the water'.

Another location in Northern Ireland associated for a time with light appearances is Lough Beg, Londonderry. Lights were repeatedly seen on Church Island in the lough in 1912. A local newspaper described the phenomenon as being

> in the form of a bright light . . . it is seldom at rest. One moment it will be observed at the foot of the tower of the ruined church, next it will flash out brightly at the tower top. Then suddenly it will wheel round and round the entire ruin at a terrific speed, shoot down to the water's edge, and almost immediately appear far out in the waters of the lake. One moment it will appear small and dim, like a candle, the next it will have the size and brilliancy of a motor lamp.[7]

Large crowds were attracted to Lough Beg to watch for the light, and as Church Island held the shrine of a saint, supernatural projections were placed on the light by locals.

WALES

Light phenomena of various kinds were intensively reported around the coastal town of Barmouth, in Gwynedd, from December 1904 to July 1905, coinciding with the local mission of Mary Jones, who represented one facet of the great religious revival going on in Wales over this period. The lights appeared in various forms – such as spheres and columns – in the immediate area around Barmouth, and particularly in the stretch of country reaching north of the town towards Harlech and including the villages of Dyffren, Llanbedr and Llanfair (Figure 7). The outbreak of lights seems to have been heralded by an arch of light, similar to an aurora borealis, which had one end resting on the sea and the other on the mountaintops about a mile inland.[8] By January 1905 the lights seem to have been regularly seen in the area by a substantial cross-section of the local population. The inci-

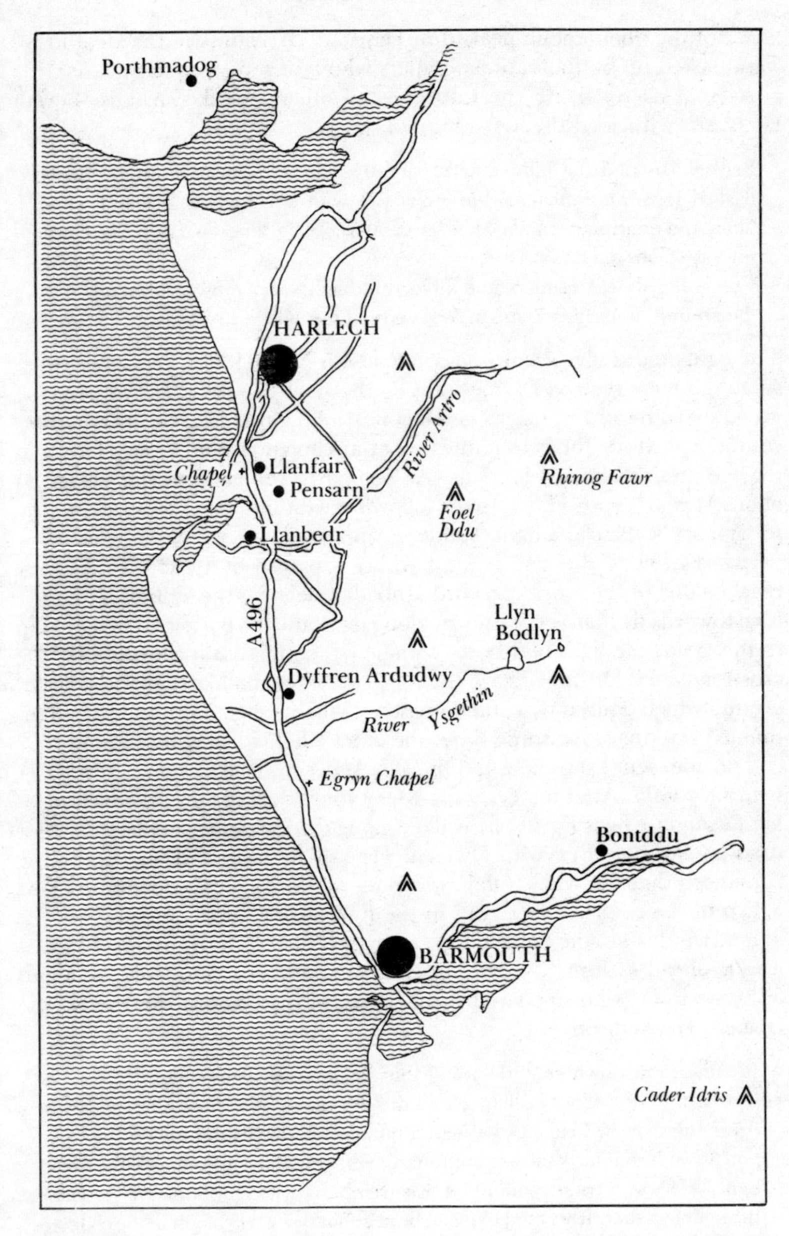

FIGURE 7 Simplified sketch map of the region between Barmouth and Harlech, northwest Wales, affected by light phenomena in 1904–5. The distance between the two towns is approximately 10 miles.

dence of the phenomena peaked in February. Accounts in the local Press were picked up by the national papers who sent reporters to the area.

As well as reporting the testimony of others, local journalist Beriah G. Evans witnessed the lights himself:

> Between us and the hills, and apparently two miles away, there suddenly flashed forth an enormous luminous star with an intensely brilliant white light, and emitting from its whole circumference dazzling sparklets like flashing rays from a diamond . . .
>
> Another short half-mile, and a blood-red light, apparently within a foot of the ground, appeared to me in the centre of the village street just before us.[9]

The lights naturally became associated with Mary Jones's revivalist mission, and were seen as Divine signs by the local populace. Evans gave the account of a Baptist minister, H. D. Jones, who, on 13 March 1905, had accompanied Mary Jones to a meeting at a schoolhouse in Ty'n-y-Drain in the mountains above Llanbedr. As her party returned to the village at about 11 p.m., with H. D. Jones following with others on foot behind the missionary's vehicle, a light 'suddenly appeared above the roadway, a few yards in front of the car, around which it played and danced'. As they reached one of the crossroads in Llanbedr (Plate 2), the light took the left turn towards the hamlet of Egryn, along the route the party had to take. Up to this point the light had been a single white light. But shortly after the crossroads, H. D. Jones saw that 'A small red ball of light appeared, around which danced two other attendant white lights. The red fire ball remained stationary for some time, the other "lights" playing around it.'

The man sent from the London *Daily Mail* saw his lights on Saturday 11 February 1905. After interviewing Mary Jones during the early evening at her farmhouse near Egryn, he walked along the Barmouth to Harlech coast road (A496) past Egryn to Dyffren. He saw nothing and was becoming convinced that the whole affair was nothing but 'a local superstition'. At 8.15 p.m. he returned, walking in the Egryn direction. He could see the lighted windows of the isolated Egryn Chapel (Plate 3) by the roadside about a mile ahead of him. Everything else was darkness. Suddenly, at about 8.20, he saw 'what appeared to be a ball of light above the roof of the chapel'. He went on:

> It came from nowhere and sprang into existence instantaneously. It had a steady, intense yellow brilliance and did not move. Not sure whether I was deceiving myself, I called to a man a hundred yards behind me on the road and asked him if he could see anything. 'Yes, yes, above the chapel, the great light.' He was a countryman and was trembling with emotion. We watched the light together. It seemed to me to be at about twice the height of the chapel, say 50 feet, and it stood out with electric vividness against the encircling hills behind. Suddenly it disappeared, having lasted about a minute and a half.

> . . . The minutes crept by, and it was twenty-five minutes to nine before I saw anything else. Then two lights flashed out, one on either side of the chapel. They seemed about 100 feet apart, and considerably higher in the air then the first one. . . . I made a rough guess they were 100 feet above the roof of the chapel. They shone out brilliantly and steadily for the space of about 30 seconds. Then they began to flicker like a defective arc lamp. They were flickering like that while one could count ten. Then they became steady again . . . they disappeared wthin a couple of seconds of each other.[10]

Although a service was going on in the chapel, Mary Jones had left for a meeting at Bontddu, on the far side of Barmouth. The reporter started to walk the four miles to Barmouth, stopping at intervals to scan the dark countryside for lights. At about 10.30 p.m. he saw a flash of light on the hill-side no more than 500 feet (153 metres) from his position. 'It shone out dazzlingly, not with a white brightness, but a deep yellow brightness,' he wrote. 'It looked a solid bulb of light six inches in diameter, and was tiring to look at.' The journalist climbed the stone wall by the roadside and made a run for the light, but it disappeared before he had covered a dozen yards. On reaching the spot where the light had been was nothing but 'bare hill-side'. Regaining the road, the reporter glanced back towards Egryn Chapel and saw another light on the road in front of it.

The *Daily Mirror* correspondent saw lights on two separate occasions. The first instance was on the same night as the events witnessed by the *Daily Mail*'s man. The *Mirror*'s correspondent had been covering Mary Jones's meeting at Bontddu, and he made arrangements to drive behind her carriage as she returned to her farmhouse. To help look for lights, the drivers of both vehicles agreed not to use their lamps. The journalist was accompanied in his carriage by a *Daily Mirror* photographer, 'a keen-witted, hard-headed Londoner'. The reporter remarked that the intimidating drive, in darkness on a narrow road between the mountains and the sea, was one he would never forget. They drove in silence for 3 miles (4.8 km) and he was giving up hope of seeing anything unusual.

> It was close on midnight, and we were nearing Barmouth, when suddenly, without the faintest warning, a soft shimmering radiance flooded the road at our feet. Immediately it spread around us, and every stick and stone within twenty yards was visible. . . . It seemed as though some large body between earth and sky had suddenly opened and emitted a flood of light from within itself. It was a little suggestive of the bursting of a firework bomb, and yet wonderfully different. Quickly as I looked up, the light was even then fading. . . . I seemed to see an oval mass of grey half open, disclosing within a kernel of light. As I looked, it closed, and everything was once again in darkness. Everyone present saw this extraordinary light.[11]

A report exists giving a similar account by one of Mary Jones' female companions. Four days later the same reporter witnessed another remarkable lightform. For several hours he had been watching Egryn Chapel along with the *Mirror*'s photographer. By 10 p.m. they had seen nothing. Then a light was seen, like 'an unusually brilliant carriage-lamp' about 500 yards (459 metres) away. The journalist rushed towards it. As he approached it, the light 'took the form of a bar of light, quite four feet wide, and of the most brilliant blue' extended across the road. 'A kind of quivering radiance', the reporter noted, 'flashed with lightning speed from one end of the bar to the other, and the whole thing disappeared. Its total duration was certainly no more than a second.' Two women behind the reporter saw the light clearly, but he found another group of people, a little further down the road on the other side of the light's location, had not seen anything at all. The photographer had his camera pointed to the chapel roof at the time the light appeared over the road, and was unable to get a shot of it.[12]

On the evening prior to the above event, on 13 February, a clergyman, hostile to the Mary Jones revivalism, saw an 'irregular mass of white light' overhead which travelled 'at lightning speed' in the direction of Egryn Chapel, whereupon

> . . . it suddenly took the shape of a solid triangle with rounded angles. I should estimate the length of the sides at five feet. Immediately over one corner of the chapel it hovered, and, in spite of the distance [about a mile] we could see every slate on the roof. The inside of the triangle sparkled and flashed as if set with a thousand diamonds. . . . For a moment, while we stared spellbound, the mystic light rested there, and then, like the lightning flashes, described an arc in the air and again settled on the opposite corner of the chapel.[13]

Another exceptional sighting was had by a local farmer who saw 'a large square of light' appear over the top of the mountains a mile from Egryn Chapel, and ½ mile (0.8 km) from his own position:

> It did not rest on the mountain-top, but was poised in mid-air about ten feet above. Between it and the mountain was a mass of white cloud. In the middle of the square was a bottle-shaped body, the bottom bright blue and the rest black. Out of the neck came a mass of fire of every conceivable colour. This . . . spreading on all sides, descended in a rainbow shower to the surface of the mountain. In less than a minute all was darkness.[14]

This account was given to the *Daily Mirror* correspondent, who found another witness who corroborated the sighting.

One of the locations furthest to the north where light events were witnessed was Llanfair. At 9.15 p.m. on 25 March 1905, a married couple were outside the chapel there (Plate 4) when they saw 'balls of light, deep red, ascending from one side of the chapel, the side which is in a field.

There was nothing in this field to cause the phenomenon.' (There is, in fact, as we shall discover in Chapter 7.) About two hours later, they saw more balls of deep red light ascend from a field high into the air. 'Two of these appeared to split up,' they reported, 'while the middle one remained unchanged.' They watched these lights for 15 minutes.[15]

Other reports tell of lights colliding together causing explosions; globes of light issuing 'radiations' (seen by a reporter from the *Daily Express*); columns of light; columns of light spewing lightballs or issuing smoky vapour; aurora-like effects, and lights in exotic shapes.

At the time of this wave of sightings there was no automatic social context for an alien craft interpretation: it was the religious link that seemed obvious to the local inhabitants. Mary Jones told the *Daily Mail* that she personally 'did not associate the lights particularly with herself', but did think they 'were Heaven-sent, and . . . connected with the revival'. When the *Guardian* asked her directly if the lights always attended her she answered, perhaps a little ruefully: 'Well, they don't go with me as much as I should like.'[16] The *Guardian* reporter went on to make perhaps the acutest observation at the time about the religious link with the lights: 'If, as is possible, the queer lights would have been seen at this time though the revival had not come, the revival would certainly have come and continued even if the lights had never been heard of.'

Rumours of lights being seen around the village of Vroncysyllte in the Dee valley between Cefn Mawr and Llangollen, about 40 miles (64 km) inland from Barmouth, led three local clergymen to conduct what we would today call a 'skywatch' on the night of 27 April 1905. The three vicars, Huw Parry, A. Lloyd-Hughes and Thomas Jones, positioned themselves on Telford's high aqueduct over the Dee close to Vroncysyllte (Plate 5). They commenced their watch at 11.30 p.m. and maintained it for over an hour. 'Twice I distinctly noticed a large ball of fire rise from the earth and suddenly burst luridly,' Huw Parry said. 'On the third occasion I saw a similar light travelling towards Vroncysyllte.' This light was also witnessed by Hughes, and all three saw two more lights afterwards. Hughes commented that 'the light resembled electricity. It rose from the earth, and was certainly not sheet lightning.'

Moving forward in time, Wales has again offered some remarkable outbreaks of light phenomena. For some months in 1977 the part of the west Wales county of Dyfed (formerly Pembrokeshire) around the St Bride's Bay–Haverfordwest–Milford Haven area was subjected to a 'flap' of sightings. Both light and dark aeroforms were seen, 'landings' occurred, and vaporous, anthropomorphic light shapes – 'silver-suited figures' – were seen appearing and disappearing. A broad cross-section of the local population saw phenomena, and there were numerous multi-witness sightings, such as the incident where a schoolyard of children and their teachers saw a strange object land, or earth itself, some distance away.

Out in St Bride's Bay is a small group of rocks known as Stack Rocks, on which bright shapes were seen moving, and to which aeroforms were reportedly seen descending. Unlike revivalist Wales, these events were interpreted – especially by the media – as visitations by extra-terrestrials. Paperback writers speculated on there being a subterranean UFO base beneath Stack Rocks. Concurrent with these events, there was a claimed outbreak of poltergeist activity on Ripperston Farm, on the southern promontory of the bay.

Another Welsh earth lights zone seems to be a band of country extending inland to almost 15 miles (24 km) south of the North Wales coastline, stretching from Colwyn Bay in the west to Shotton in the east, covering parts of the counties of Clwyd and Gwynedd (Figure 8). There has been outbreaks of odd lightball and similar events peripheral to this area over the years, but within the 'window' itself there has been more activity, including one of Britain's most celebrated close encounter cases.

An example of one of the well attested earlier cases within the region took place at Brookhouse Mill, on the A525 immediately southeast of Denbigh, on the night of 26 January 1979.[17] Keith Jones, then a chef at the Brookhouse Mill restaurant, was crossing the road to the old mill housing the restaurant at 8.45 p.m. when he noticed a curious illumination occuring around him. He looked up to see an elongated oval object emitting a piercing white light from a dome-shaped protuberance on its underside. The thing was no higher than 150 feet (46 metres), yet made no more sound than what seemed to be the rushing of displaced air as it travelled. The light was flashing at about one-second intervals, and the object was moving fast. Although it was a clear, frosty night, Jones could not make out any structural details on the object other than an impression that the object's surface was an aluminium colour. He immediately went inside the restaurant to get witnesses, and his employer and a customer came out to watch. The object was by now further down the valley, but had slowed down its rapid flight. It proceeded to flit from one position to another 'like a dragon-fly over water' while the three men watched. The police were called as the light moved off and disappeared in the distance after a display lasting around 15 minutes. The two experienced police officers who arrived to question the witnesses had themselves seen the brilliant flashing light, and one of the policeman had, in fact, witnessed a strange light the previous evening in the general area. Keith Jones, who had seen a similar light some 18 months earlier a few miles further south at Llanynys, was later questioned on the telephone about the Brookhouse Mill incident by an intelligence officer from RAF Valley, on Anglesey. He asked Jones if there was any water or waterfalls, waste land or power lines in the vicinity of where the sighting took place. These questions suggest that the authorities had some understanding of the phenomenon they were dealing with even as far back as 1979, as will become more apparent in Part 3.

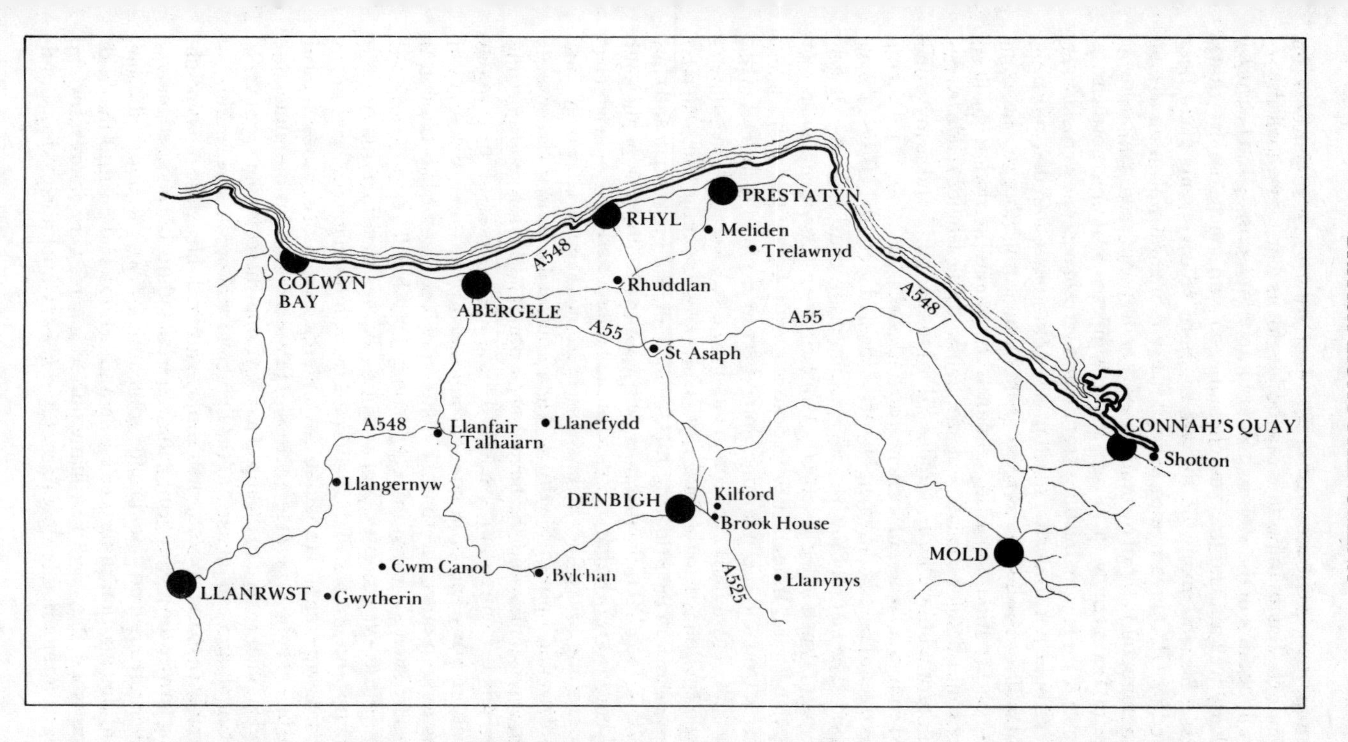

FIGURE 8 Simplified sketch map of the area of North Wales subject to outbreaks of light phenomena activity. Some particular locations mentioned in the text are marked. The distance between Prestatyn and Denbigh is roughly 10 miles.

An important event occurred late on 14 April 1984, between Llanrwst and Denbigh. It is an important case because of its date, as we shall also see in Part 3, because of the intense military interest it aroused, and because it is a clearly well attested unusual phenomenon with some close eyewitness reports. This and many of the area's more recent events have been reported by UFO investigator Margaret Fry.[18] She was woken up at half-past midnight on 15 April by much activity at the police station opposite her cottage in Llangernyw (a village about 6 miles – 9.6 km – northeast of Llanrwst). People were bustling around and vehicle beacons were flashing, but Fry could not understand what was going on. Later that morning, around 6.30 a.m., she was awoken again by an army helicopter hovering by the police station depositing a man who entered the station. Military activity continued for the following six hours: the helicopter made trips to the hills around Gwytherin, 4 miles (6 km) to the south, and army trucks arrived with small teams of soldiers with backpacks. At 7 a.m. Fry went out on the street to be asked by neighbours if she had seen 'the UFOs landing last night'. The fact that she had not was, naturally, cause for some chagrin on the part of such a dedicated UFO investigator! Many people had seen the lights because, though they appeared shortly before midnight, it was lambing season and sheep farmers and others were out in their fields. Two orange balls of light had been seen in the sky north of Llanrwst, moving in a slowly descending arc to the southeast. They passed over the Cledwen river and came down in the fields around Cwm Canol near Gwytherin and close to the B5384 road. This course was clearly visible to anyone looking south from Llangernyw. Margaret Fry eventually questioned 20 people who had seen something of the events, but most were not keen to talk, or to have their information put into writing. These included a local social security worker, a policeman and the farmer at Cwm Canol. Nine venture scouts camped on the farmer's land likewise refused to give details. It appeared to Fry that the army had been going around advising witnesses close to the landing to keep quiet and that the lights were 'just flares'. It is difficult to understand such a buzz of military activity for a few flares.

One of the eyewitnesses prepared to talk was a farmer just outside Gwytherin who saw one of the lights pass by within feet of him as it came to rest near Cwm Canol. He said it was an 'eerie experience'. It was a pink-orange ball of light 'several feet across'. He watched it for two minutes as it descended, then rushed to the top of a ridge from where he could see over to Cwm Canol, and caught sight of the lightball for about another minute. It was very close to the ground, then simply vanished. He spoke to his neighbouring farmer the next morning; he had seen nothing but claimed that the army was on his land 'looking for something'. About a mile and a half north of this location, a farmer's wife and her son had been in the fields lambing when they saw a brilliant purple sphere, very fluorescent and moving at a remarkably slow pace. Over Cwm Canol it exploded soundlessly, producing a shower of sparks out of which emerged a white disc. This then descended vertically and was lost to view.

All witnesses were well used to whatever military activity there was in the area, and these events fitted nothing they had experienced previously.

Strange cigar-shapes and lights had been seen earlier in 1984 out to sea off the coast near Colwyn Bay and Abergele, and in July 1984 a retired *Times* journalist at Llanfair Talhaiarn, four miles northeast of Llangernyw, saw a bullet-shaped object 'salmon pink, like a shrimp' hovering over an old oak tree before rising vertically and disappearing. Also in July, around the fifth, a woman saw a domed object on the ground in daylight near Llanfair Talhaiarn, a village between Llangernyw and Abergele. It had what appeared to be windows, and was the colour of yellow stone. It looked, she said, like a 'Moslem mosque'. There is no known building near the location to account for the object. This month was also significant seismologically, as we shall see.

In late 1985 early 1986 the area experienced another 'wave', peaking in January. In late August 1985 a motorcyclist on the minor road between Llanefydd and Llanfair Talhaiarn encountered at twilight what seems to be a domed object identical to that which the woman had witnessed over a year earlier, and at about the same place. The man, a local, was unnerved by the unfamiliar object. Two witnesses in separate locations saw a 'yellow egg' rise from the Denbigh Moors near Bylchau in November; three witnesses saw a yellow-white object execute an arc in the sky at Shotton, Deeside, in December, and later that month two women saw a yellow-green light emerge from a quarry at Meliden, near Rhyl. This rose up into the clouds, then reappeared to descend and land near the A547 road. On landing it changed to a brilliant white globe with 'a red rotation to it'. Later that same night other witnesses at Meliden saw the sky light up and a disc with multi-coloured spinning stripes appear and fly close to the edge of the Meliden hills. On 3 January 1986, a family of five saw a strange light hovering around the TV transmitter mast on Moel Parc mountain near Denbigh. The light was a hazy red at the bottom, and white at the top. After hovering stationary for a time, it suddenly flew directly over the witnesses, displaying a rotating ring of coloured lights on its underside. In the early hours of the following morning a local government officer saw from her car a red oval light descending in a gap between the mountains of Coed Cwm and Moel Maenfa while travelling along the A547 towards Rhuddlan. She saw it land and light up the ground around it. Another red oval was seen in the evening of that day, glowing in a field, by a woman at Trelawnyd, about 2 miles (3 km) from the previous location. It went 'smaller and smaller and became a dot before disappearing', the woman stated. Later that evening, four of her neighbours also saw an oval object with a red central light flanked by two flashing white ones. This object was hovering above electricity lines on farmland. On 10 January 20 people at

Prestatyn saw red lights changing shape over Meliden a mile or so to the south. These eventually 'faded away'. At 9 p.m. on 22 January two adults and three children saw an object like three saucers stuck together over Prestatyn which then flew south to the Meliden hills where it performed aerobatics, finally speeding westwards towards Snowdonia. January ended with a taxi driver at Connah's Quay, Deeside, seeing a dull metallic-looking object with a bright blue light at rear, three strips of 'frosted' white light underneath and a ring of white lights around its circumference, hovering low near a power station. Other people also saw the object, and Margaret Fry was able to obtain three independent sketches from witnesses, all showing substantially the same object.

Further events were reported up until April, with dramatic cases from Shotton, Flint and Prestatyn. Activity started up again on 23 August 1986 with sightings at Mold, Ruthin, Llanfair Talhaiarn and Rhyl, on the coast. In October further incidents took place in Prestatyn, including a reported poltergeist outbreak that made the national TV news. Since 1986 aerial phenomena reports have been more sporadic.

The difficulty encountered in identifying some earth lights haunts is exemplified by the discovery of light phenomena in the sparsely-populated heart of Wales, sometimes called 'the green desert', to the west of Rhyader and Llanidloes. Here the general lack of potential witnesses is compounded by the usually quiet acceptance of the phenomena and the reluctance to talk about them on the part of the limited number of people who do live in the district. I live barely 50 miles (80 km) from the area, yet I got to hear about light events there only because of two chance occurrences.

The first of these was an event which broke the usual silence on the local light phenomena. I obtained information through Phil Rickman of BBC Wales. He has written the following account for this book:

> On 19 November 1987, BBC Wales sent me to a public meeting in the town of Llanidloes held to discuss the activities of British Aerospace deep in the Hafren Forest.
>
> It was a night of torrential rain, but the meeting had attracted over fifty people – a big turnout for a town of fewer than 2500 people.
>
> Protests, led by the environmental pressure group Greenpeace, had been voiced over British Aerospace's refusal to disclose the nature of research being carried out in a former quarry in the heart of the forest. There were fears – rejected by British Aerospace – that the testing site was being used for something environmentally damaging. The protests were supported initially by the local MP, who later declared that British Aerospace had satisfied him that nothing dangerous was going on in the forest.
>
> However, at the public meeting in Llanidloes, several people spoke of seeing curious strobe-type lights on the tops of hills around the forest. Most of these people were comparative newcomers to the area who had immediately

made connections between the lights and whatever was going on at the British Aerospace testing site. I recorded interviews with these people and then talked to a local farmer at the meeting. Tecwyn Hamer had lived in the area all his life and was among those worried about what might be taking place at the British Aerospace site. I asked him if he had seen the lights. He said he had – but he did not connect them with British Aerospace.

'I've seen them since I've been a child,' he said. And it wasn't just in the forest – he had seen them over the brows of several hills in the area. 'Strobing lights.'

Q: 'What do you think it is?'

A: 'I haven't got a clue.'

Q: 'It doesn't appear to be any sort of natural phenomenon, does it?'

A: 'No, not really. No.'

Q: 'And it's not traffic?'

A: 'Oh, no, it wouldn't be traffic. I just can't explain what they are. When I was a child I was always told they were, you know, something like the Northern Lights. But as far as what's going on in that quarry, I just don't think it's tied in with the lights at all.'

spoke to Mr Hamer myself, and he confirmed that he had seen the lights rom time to time over many years, and did not think they were to do with he secret activities of British Aerospace (who said they were in the area ecause it was far from busy domestic and industrial electromagnetic ctivity). The lights appeared as flashes over the crest of hills all around Iafren Forest, which itself rises above the picturesque Llyn Clywedog eservoir, and is situated by the eastern flank of the Plynlimon massif near ne sources of the rivers Wye and Severn.

This appearance of the lights was similar to that described by another vitness I spoke to, Susan Wood of the Llanidloes area. She described them s like 'a quick electric storm' in the sky, with the flashes like 'silent sheet ghtning but much more brilliant . . . they are in the sky but reflected on ne hills'. The colour of the flashes ranged from 'white to violet-indigo'. 'eople had seen the lights from Staylittle, a tiny, isolated village at the north nd of the reservoir, around the old lead mines at Fan on the east side of lyn Clywedog, and in the Dylife area a few miles northwest of the reservoir. There had been an increase in this luminous activity in late summer nd autumn of 1987. The flashes had occurred over the same period as curous subterranean rumblings, which more than one person described as ounding like 'giant garage doors opening and closing underground'.

Dylife ('Place of Floods') was mined for lead in Roman times, and was amous as a mining hamlet up until relatively recent times. Strange lights ave been seen not only above ground at Dylife, but also beneath the surace. This information came as a result of contact between journalist Keith tevenson, who goes by the pen-name 'Llowarch', and a man we shall call

'Jeff'. The account was given by Llowarch in his 'Weird Wonders of Wales' column in the *Cambrian News* (14 November 1986). In 1984 Jeff was exploring a disused Dylife mine, and while making slow progress through one level, came across a pool of water above which was what he felt could be a natural fissure in the rock roof. He suddenly heard a humming 'like somebody's voice' from the other side of the pool. He could see nothing in the beam of his flashlight as he peered ahead. He switched off his torch, in case he could see another person's lamp. About 10 yards (9 metres) ahead of him, Jeff saw 'a white or pale blue shape about the size of a small man. It gave off a sort of glow, but not like a torch.' Jeff made a hasty retreat from the old mine. The evening, he was talking to locals about the former mine workings, without mentioning his experience of earlier in the day. One local, in his late fifties, remarked spontaneously that he had seen lights on more than one occasion emerge from mine levels at night. These would 'rise into the air and move off through the sky, but not quickly.'

We saw in the previous chapter that there used to be a tradition amongst early miners that ore veins could be located by watching for lights emerging from the ground. This is further augmented by lights seen on the hills around the Tregaron district on the western extremity of 'the green desert' area. I asked Llowarch, who lives near Tregaron, if he had heard of lights being seen in his area. Not only had he heard about them, he had seen them himself! Blue lights would occasionally be seen popping out of hillsides. Locals took them for granted, it seemed. There were fairy legends associated with them, but another tradition was that there were minerals in the areas where the lights appeared. The old memory lingers on, apparently. Llowarch was frankly amazed that these lights could be the subject of controversy, when they were such accepted phenomena in his locality. This association between light phenomena and minerals or mines will be a recurring theme in this book.

Another kind of light haunts Tregaron Bog (Cors Caron) to the north of the town and to the west of the B4343 road. This had been the site of a reported UFO landing in the 1950s, and lights are occasionally seen on the bog. Prompted by journalistic interest, Llowarch and his wife carried out a watch one June night, and succeeded in seeing 'a long, low light moving across the face of the bog. . . . It glowed dull red. The light was northwest of us, travelling southwards. A swishing noise came from its direction.'[19] This may be dismissed as 'Will-o'-the-Wisp' by sceptics, but it is clearly an unknown light phenomenon. On another occasion, a woman travelling at night along the B4343 saw a light 'of great brightness' moving across the bog, almost at ground level. It had a blue tinge.

About 12 miles (19 km) due east of Tregaron is the Elan Valley, an exceedingly remote spot, peopled only by scattered sheep farmers. The valley, only 15 miles (24 km) south of the Hafren Forest area, contains a string of reservoirs which reach down almost to Rhyader. Shortly before hearing about the Hafren lights I was told, quite by chance, that certain sheep farmers had seen lights for many years in particular locations around the valley – one of these being a quarry next to a waterfall close by the Caban Coch reservoir. People are not keen to talk about these events, and further enquiries are being made at the time of this writing.

So, it is quite clear that even from the zones we know about, Wales presents a striking set of earth lights locations for its relatively small geographical size. I believe this is due to its fascinating geology and interesting seismic history. But we shall return to these aspects in Part 3.

ENGLAND

When Paul McCartney and I studied UFO distribution in England and Wales for *Earth Lights*, weighting 800 reported sightings against population, Cumbria, or the Lake District, in northwest England, showed up as something of a window area. There has been at least one recorded flap there as well as occasional reported phenomena over long periods of time. But the aspect of earth lights I want to concentrate on for this region relates to phenomena reported around the Castlerigg stone circle, as it indicates a possibly very special facet of the earth light enigma.

Writing in *English Mechanic and World of Science* (17 October 1919), T. Sington recounted an incident that had occurred 'during Easter' some years earlier. He had been climbing Helvellyn with a hotel acquaintance. They were returning to Keswick one dark evening, passing close to Castlerigg, magnificently situated a mile or two above the town (Plate 6):

> We all at once saw a rapidly moving light as bright as the acetylene lamp of a bicycle, and we instinctively stepped to the road boundary wall to make way for it, but nothing came. As a matter of fact, the light travelled at right angles to the road, say 20 feet above our level, possibly 200 yards or so away. It was a white light, and having crossed the road it suddenly disappeared. Whether it went out or passed behind an obstruction it is impossible to say. . . . .
>
> We then saw a number of lights possibly a third of a mile or more away, directly in the direction of the Druidical circle. . . . Whilst we were watching a remarkable incident happened – one of the lights, and only one, came straight to the spot where we were standing; at first very faint, as it approached the light increased in intensity. When it came quite near I was in doubt whether I should stoop below the boundary wall as the light would pass directly over our heads. But when it came close to the wall it slowed down, stopped, quivered, and slowly went out, as if the matter producing the light had become exhausted. It was globular, white, with a nucleus possibly six feet or so in diameter, and just high enough above ground to pass over our heads.

Sington noted that the lights moved horizontally near the stone circle in opposite directions to one another at times, indicating that their motion was not due to wind. The experience made him wonder if the builders of the circle had seen the lights in the area from time to time, due to 'some local conditions at present unknown', and had built the megalithic structure out of awe for what they interpreted as spirits or gods.

It is an unanswered question, but one that is currently being considered by 'Earth Mysteries' researchers in their investigations into the geophysical properties of prehistoric stone sites. I suggested in *Earth Lights* that as Paul McCartney had confirmed stone circle distribution favouring areas adjacent to surface faulting, the higher incidence of lights in such landscapes could possibly have been a factor in prehistoric shamanistic practice, the situations of stone circles being associated with this as places where spirits manifested. This suggestion was criticised or dismissed by some. Sington's account, however, had not been uncovered when *Earth Lights* was written, so perhaps the suggestion now takes on a little more credibility. Furthermore, several other accounts of light phenomena being observed at or near megalithic sites have come to my notice. This aspect of light phenomena is more fully explored elsewhere.[20]

Linley is an isolated hamlet about 9 miles (14 km) south of Telford. When Jonathan Mullard was researching the county of Shropshire he came across an old account of curious lights haunting Linley around 1913.[21] The hamlet is comprised of a Norman church, sixteenth-century Linley Hall, and a few adjacent cottages. A track links the church and hall (Plate 7).

In the outbreak, fiery balls of light were seen to form on the top of the church tower and roll down its sides. Lightballs were seen flitting around one of the cottages, and there is a suggestion they also appeared inside Linley Hall. Strange noises were heard, including explosions in the air. In addition to the lights, light and dark spirals of 'vapour' appeared. Curious side-effects also took place: door latches would stick mysteriously as if magnetised, and poltergeist phenomena occurred – teacups, furniture and garments were seen moving around of their own accord. It seems this outbreak of strange phenomena lasted for several weeks, then ceased entirely.

A curious episode of strange lights developed in Norfolk over 1907 and 1908. Naturalist R. J. W. Purdy first saw one of the lights one February night in 1907, near the village of Foulsham, between Fakenham and Norwich. He and his son saw a luminosity moving 'up and down vertically with great rapidity upwards to a height of some 50 or 60 feet'. This manoeuvre was repeated a number of times, and when it swept close to the ground it sometimes 'moved horizontally very quickly for about 100 yards, and then back'. The witnesses observed the light for 20 minutes and said it looked like a carriage lamp but was tinted red in the centre. Purdy saw a similar light again in December of that year, in the same locale. He called it to the attention of some nearby countryfolk who said it 'looked funny' and had never seen anything like it. He went to get members of his family and some servants and they watched the light's antics for a short time. Then he and a labourer went down to the field where the light was flitting around, but it moved off as they approached. Purdy had a telescope with him through which the light looked like 'a large lamp surrounded by mist'. It suddenly disappeared. Perhaps because of his naturalist background, and because the movements of the light suggested bird-like motion to him, Purdy became convinced that the phenomenon was a luminous owl. He suggested that the bird may have become luminous through fungal disease or resulting from its contact with phosphorescent wood in its journeys or in its nesting place.

As more and more people saw the lights, which sometimes cavorted in pairs and at distances up to 30 miles (48 km) from the Foulsham area, the idea of luminous owls was taken up by several other commentators. But natural phosphorescence, which has a muted glow, simply could not account for the intensity of some of the lights that were reported – one was stated to be as bright as 'a several candle-powered arc-lamp' and lights on more than one occasion were seen to light up whole trees.

Whatever the nature of the lights, all that observers ever actually saw *were* lights, and no luminous owl has ever been studied by science. The idea of glowing birds is merely another conceptual frame for mystery lights – it at least makes a change from dragons, ghosts or alien spaceships.

The last light in 1908 was reported in May, but it was seen once more in February 1909.

The farm labourer John Clare, who became known as 'the Peasant Poet', wrote of unusual lights around the village of Helpston, Cambridgeshire, between Peterborough and Stamford, in his *The Journals, Essays, and the Journey from Essex* in 1830. There was 'a great upstir' about the lights, but it was only when Clare saw some for himself that he felt 'robbed . . . of the little philosophic reasoning which I had about them'. The first light he saw was over Eastwell Green, and he initially took it to be a meteor, but as it became larger and began moving up and down he observed with increased interest. He then saw another light 'rising in the southeast' over Deadmoor. The two lights moved towards each other, and they danced and chased around until 'they mingled into one in a moment'. Clare specifically noted that the lights flew against stiff winds at times, and found it difficult to think of them as 'vapours'. Clare talked to one village woman who had seen up to 15 lights moving around over Eastwell Moor at the same time – 'there is a great many there' he was told. Clare saw several more subsequently, and one night a light approached him as he was walking between Ashton and Helpston. Clare's account of this is given here in his own inimitable style:

> it came on steadily . . . & when it got near me . . . I thought it made a sudden stop as if to listen to me . . . it blazed out like a whisp of straw and made a

> crackling noise like straw burning which soon convinced me of its visit the luminous haloo that spread from it was of a mysterious terrific hue & the enlarged size & whiteness of my own hands frit me . . . the bushes seemed to be climbing the sky every thing was extorted out of its own figure & magnified the darkness all around seemed to form a circular black wall. . . . I held fast by the stilepost till it darted away when I took to my heels & got home as fast as I could.

The setting for another intense outbreak of light phenomena, in the 1920s, was the Burton Dassett hills which lie to the northwest of Banbury, about 10 miles (16 km) south of Warwick.[22,23] They stand as detached outcrops of the main Cotswold range, and rise out of an otherwise flat plain commanding views in all directions. Ironstone quarrying has taken place on the hills. Burton Dassett village suffered the ravages of the Black Death and tenant evictions centuries ago, and all that exists today is the beautiful All Saints' church (Plate 8) alongside a scattering of cottages and farms. The present church dates from the twelfth century, but one certainly existed earlier at the same site. The main claim to fame of the Burton Dassett hills is that they overlook the valley where the Battle of Edgehill took place between Roundheads and Cavaliers on 23 October 1642. Many ghost tales are associated with the area, including the widely witnessed appearance of ghostly armies materialising in mid-air re-enacting the Battle of Edgehill two months after it had taken place.

The lights seem to have first been talked about in December 1922, and to have centred on Burton Dassett churchyard, close to which is a holy well. The first report came from William Neale, an 'aged cowman', who lived with his wife in a cottage on the hillside just below the church. He claimed to have seen 'the' light 'thousands' of times, moving around the churchyard, floating over the hills between Burton Dassett and nearby Fenny Compton, and soaring over a derelict farmhouse near his cottage. Although reporters and writers who interviewed Neale came away convinced that he was an honest sort, local people would not at first believe him. So parties of villagers ventured out late at night to see if they could catch a glimpse of the mystery light.

One local sceptic, George White, set out onto the hills with two companions early in February 1923. He did so out of 'sheer curiosity' because he 'frankly did not believe a word' of Neale's story. Here, in White's own words, is what transpired during the skywatch on the hills:

> It was about seven o'clock, and when we had been there a short time Mr Shearsby said 'Here it is!' We turned round, and about 200 yards away was a strong and dazzling light. . . . It was a perfectly lovely sight, and it held us fascinated. It flitted about and passed through bushes and over fences at a great speed; then, with a final flash it disappeared, we know not where. We could feel it hover around, and it appeared to be looking for something, for the light

swept the ground. I had my field glasses and was able to get a 'close-up' view. It was a kind of reddy-blue mixed, but beautifully blended. Later, when we saw it round Burton Dassett Church, there was a tinge of orange colour. There was nothing whatever to be afraid of, and I have decided to go again, hoping to see it![24]

Numerous other local sightings were made during the same period, and as news of the lights reached the national press, Burton Dassett became the focal point for the curious from all over the country. A reporter from the *Birmingham Post* saw the light from the summit of the highest hill in the neighbourhood on 16 February:

> It appeared to be fully a mile away, but its radiance was such that the sky was faintly illuminated for several miles. . . . The light itself was steady and vivid, but the wide beams which were thrown upwards flickered like a failing lamp, and the general aspect of the phenomenon was not unlike the dying rays of the Aurora Borealis. The light appeared to be travelling towards Fenny Compton, and, as usual, by the time it had reached a woodland tract known locally as Bottom Cover, it faded from view . . . .
>
> The most remarkable feature of the display is the form and intensity of the glow . . . its appearance is not likely to cause more than a flutter in the scientific world, but skilled investigation might possibly be well repaid.[25]

It is a sad reflection on the state of our scientific curiosity that such investigation, over 60 years later, has barely begun.

A correspondent from the *Birmingham Gazette*, in the company of other 'ghost hunters', was the next journalist to see the lights, a few days later from the vantage point of the derelict farmhouse around which William Neale had seen the lights cavorting:

> We were facing the Edge Hills across the valley, and the light, well-defined and spherical, moved across our field of vision with its peculiar switchback motion from left to right, disappearing as suddenly as it had come. A few moments later we saw it again, a pin-point of light, which seemed to be growing larger every second, moving, however, hardly at all.
>
> 'It's coming at us,' gasped a nervous ghost hunter, and come it did until, like a dull yellow eye, it was glowering at us apparently from beneath a nearby tree.
>
> There was a concerted rush to the spot . . . when we were at the foot of the tree . . . there was nothing at all – nothing anywhere except a derisive light, circling at the side of the ruin we had just left. After that we saw it again three or four times in every quarter of the compass and flitting in every direction.[26]

The same correspondent saw the lights on another evening, also. His overall observations on the phenomena are of great value; he saw a light move over the hills against 'quite a gale'; the lights usually floated 'anything from two to thirty feet' above the ground; the typical motion of a light was 'undulatory, with sudden dives downwards, at which it generally disappears, to appear again perhaps five hundred yards away'. The journalist stated that the Burton Dassett light was 'a well-defined light, spherical in shape and without radiations . . . [it] glows with a dull yellow colour, tinged when seen near at hand, with red. To liken it to a Chinese lantern describes it exactly.'

The *Birmingham Gazette* journalist also commented that 'neither the appearance of the light nor its habits or habitat' accorded with a 'Will-o'-the-Wisp' explanation. He dismissed a 'luminous owl' interpretation, too, because 'no owl could give off so brilliant a light. If he could, would the phosphorescence last for his nightly appearances during a whole winter . . .?'

It was reported that the older residents of the Fenny Compton district 'accept the appearances of the "ghostlight" as quite a commonplace occurrence, and it is claimed that lights similar . . . have been seen over the Burton Hills in damp weather for over ninety years.'[27]

But the appearance of the light seemed to cease over the following months, and no more was heard of it for the rest of the year. It put on a final, particularly brilliant, show on 26 January 1924 – a significant day as we shall learn in Part 3.

From at least the turn of the century for a decade or two, an area between Chipping Norton and Burford in Oxfordshire had a light phenomenon that was known locally simply as 'The Light' – a refreshingly uncomplicated context. The exact location is impossible to determine from the sparse records available, but it seems to have haunted a road along the crest of some hills near the village of Shipton-under-Wychwood. By 1919 it had been 'going on for many years'. It sometimes hovered over treetops near the road, while at other times it could be seen moving rapidly along the road 3 or 4 feet (1 metre) above the ground:

> From a distance, it might easily be mistaken for a motor-car headlight, but an eye-witness asserts that it is really like a ball of 'blood-red fire'. Another witness tells of his encounter with 'The Light'. . . . He was going up the hill one night when he saw what appeared to be a shepherd'a lantern being carried down the hill towards him, the illusion was aided by the fact that the light was swinging in just such a way as would a lantern when carried. Just before it reached him, however, it turned from the road. As he came level with it, he naturally remarked 'Good night, shepherd!' There was no response, however, and as he watched, the light drifted (drifted is his word) up and up until, some 200 feet from the ground, it disappeared.[28]

Sometimes it was claimed that there were two lights moving in unison, 'as if they were at either end of a pivoted lever'.

A correspondent to the *Royal Meteorological Magazine* (1891) gave an account of light phenomena around the town of Crowborough, situated south of London between Tunbridge Wells and Brighton, and adjacent to Ashdown Forest in Sussex. The writer had 'for some years now . . . in common with many others' been seeing light phenomena on an estate at Crowborough, playing around in the evening at various times, singly and in pairs:

> It does not dance about, but now and then takes a graceful sweep, now to quite a height, then makes a gentle curve downwards, after sparkling and scintillating away for ten minutes or more.
>
> In winter there is generally a very remarkable one which appears to rise in Ashdown Forest beyond this estate, which I have known to keep steadily in the air for half-an-hour, and then sail away a long space and stop again.

The writer was convinced that the lights foretold of unsettled weather.

An old farm road near Padworth, Berkshire, along which sheep and cattle were taken to the market at Reading, several miles to the northeast, was an apparent earth lights location. Edward Hobbs, the parish clerk from 1871 to 1911, often reported seeing lights along that road at certain times of night. They resembled two yellowish lamps and appeared to travel quickly. When fairly near, they 'just disappeared, and a strange uncanny stillness enveloped the watcher'. *The Berkshire Village Book* (Berkshire Federation of Women's Institutes, 1972) seems to hint that the lights are still seen from time to time.

In the latter part of the nineteenth century, Mere Down, to the northeast of the small town of Mere, Wiltshire, was well known locally as a haunt of curious lights. These were often referred to as 'Kitty Candlestick' – the marsh gas explanation prevalent at the time. But the delightful description of an encounter with a Mere Down light by one old country fellow tells us that the Will-o'-the-Wisp explanation is untenable:

> It went across the Down like a flash of lightning. By-and-by it came back again, and we looked at one another a bit, and I said, 'What! have 'ee a-lost your way?' and off he went again. It was a beautiful light as big a plate.[29]

It is probably worth remarking that Mere Down is barely 9 miles (14.5 km) from Warminster, and well within the compass of the 'UFO' phenomena that plagued that area in the 1960s.

Between Ashburton, below the eastern rim of that wild, granite upland known as Dartmoor, and Princetown, near the centre of the moor, lies the village of Hexworthy. During the summer of 1915, unusual floating lights were seen in this area of Dartmoor that became the concern of the Naval Intelligence Department, as they feared the lights might signify the activity of German spies. In the event, nothing was ever found to support this, and it would be difficult to understand what German spies might be hoping to accomplish by flashing lights on a remote moorland in any case. But because of this military interest, we do have a few eyewitness reports of the phenomena – even one by the main investigating intelligence officer.

The intelligence officers noted but largely ignored early reports of lights on Dartmoor – until they received reports from the hamlet of Sherril, close to Hexworthy. Here, a woman and her daughter had seen 'on several occasions' a bright white light rise from a point a few hundred yards to the east of the Hexworthy tin mine on Down Ridge. The light would swing westwards across the valley for several hundred yards, until it was west of the mine, and then disappear. Sometimes it was seen against the sky, at others against the bulk of Down Ridge. It always rose from the same spot. The investigating officers found both the womens' accounts to corroborate each other, and further noted in their report of December 1915 that 'this floating light against Down Ridge has been reported from the Hexworthy district on several occasions since, the last being a few nights ago'. As they had interviewed the women in July, it is reasonable to assume the light phenomena occurred in the area for at least six months.

By the middle of August, reports of floating lights were coming in from several persons in an area around Dartington Manor near Totnes, about 4 miles (6 km) southeast of Ashburton. Both the navy and the police questioned some of the witnesses, and became convinced that an actual phenomenon was being described. So on 4 September, Lieutenant Colonel W. P. Drury obtained permission from his superiors to spend a night looking for the light. He stationed himself with other witnesses near the Dartington sight point:

> About 9.30 that night . . . we observed a bright white light, considerably larger in appearance than a planet, steadily ascend from the meadow to an approximate height of 50 or 60 feet. It then swung for a hundred yards or so to the left, and suddenly vanished. Its course was clearly visible against the dark background of wood and hill. . . . We were within a mile of the light and both saw its ascension and transit distinctly.[30]

The military took these lights seriously, as can be gathered from the following memo from the Commander-in-Chief's office, Devonport, dated March 31 1916:

> Countless reports have been received by the Military Intelligence Department of 'Floating Lights' on Dartmoor from many independent sources, but the description is nearly always identical. A ball of light is seen to rise perpendicularly from the ground to a height of anything from 30 to 60 feet. It lasts varying times from 2 to 25 minutes. All efforts (and there have been many) to capture these lights, or to detect the originators thereof, have hitherto failed, although it seems evident that the places of origin are comparatively constant.[31]

There are hints that the Ashburton area was prone to lights prior to the 1915 happenings, lights that are not amenable to a marsh gas explanation. For example, the Victorian antiquary S. Baring-Gould recalled some lights observed by a group of witnesses living close to Ashburton during the autumn of 1898. They saw a light 'of a phosphorescent nature' in meadows between Ashburton and Pridhamsleigh, about a mile to the south. 'It appeared to hover a little above the ground and dance to and fro,' Baring-Gould wrote, 'then race off in another direction, as if affected by currents of air. This was watched during several evenings.'[32] Although prone to interpret all unusual lights as Will-o'-Wisp, the antiquary had to admit that the area in question, while formerly a marsh, had been 'drained perhaps sixty years ago'. Baring-Gould insisted that 'implicit reliance' could be placed on the account of the principal witness. The same man had also seen a 'similar flame in the form of a ball some forty years previously' in the same general area. In 1899 Charles Purnell mentioned in a despatch to the science journal *Nature* that he had 'witnessed several auroral displays at Ashburton'.

Odd light effects are still noted from time to time on the moor. On 1 September 1981, artist Chris Castle, known to me personally, saw a curious phenomenon from the area around Drizzlecombe, a remote place with many standing stones, about 6 miles (9.5 km) southwest across the moor from Hexworthy. In sunny conditions, early in the morning, something caught Castle's attention:

> Out of the corner of my eye, just above the distant western horizon, I saw a glitter of brilliant light. As I turned to look I just saw what looked like a cluster of mirrors falling down behind the ridge. I was astounded. 'What was that?' I thought, and at that, as if in answer to my question, it happened again – only this time I was able to observe it more clearly. The same thing. Crystal clear. As if someone had thrown up a pack of cards in the air, a cluster of square or

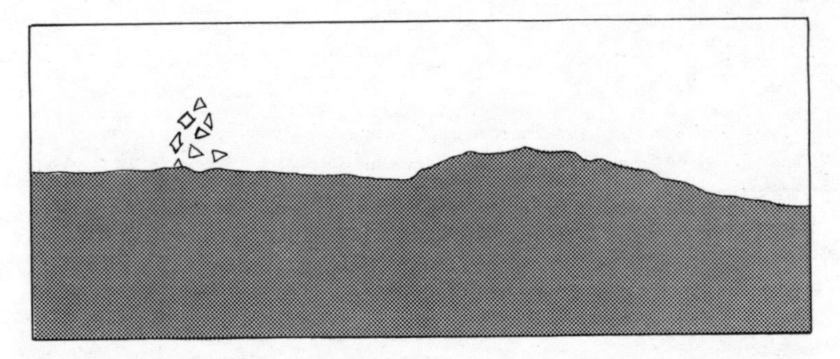

FIGURE 9 A depiction of the light phenomena seen by artist Chris Castle on Dartmoor. (*After a sketch by C. Castle.*)

4 PROJECT PENNINE

Project Pennine was founded in 1986 by David Clarke and Andy Roberts, both experienced UFO investigators and editors of specialist magazines. They received valuable assistance from Philip Mantle, Martin Daglass, Philip Shaw (of Glossop Mountain Rescue Team), Tony Dodd, Granville Oldroyd, Paul Bennett, Jenny Randles and David Kelly. Clarke and Roberts were aware of the failure of ufology to research into the areas where UFOs were seen, particularly over long periods of time. They also felt there was too much concentration on the tiny proportion of 'high strangeness' cases at the expense of the more consistent lightball phenomena. They therefore devised a research programme to help mitigate this situation. As it was already apparent in standard UFO reports collected over previous years that the chain of hills in northern England called the Pennines had provided more than its share of sightings, and as they lived locally, this was an obvious area on which to base their investigations.

They began collecting and collating data relating to anomalous light phenomena from records of contemporary sightings, folklore and historical sources. The final objective is to compile an accurate, comprehensive catalogue of unusual lights in the Pennines which can be freely accessed by researchers and also made available to the scientific community. In planning this kind of work, they were developing the idea of small-scale regional studies pioneered in *Earth Lights*. Project Pennine is still ongoing, but is proving so spectacularly successful that enough material is at hand to make this first, if selective, publication of results to date appropriate.

THE PENNINES

This chain of hills extends from the Cheviots on the Scottish border in the north, down through the Yorkshire Fells to the Peak District of Derbyshire in the south, ultimately merging with the Midlands Plain around the valley of the river Trent. It is the watershed of the main rivers of northern England and provides a ribbon of desolate wilderness wedged between urban areas such as the Manchester conurbation to the west and Sheffield,

Bradford and Leeds to the east. In all, the Pennines cover several thousand square miles and form a dissected plateau with summits ranging up to a little under 3000 feet (917 metres) high in the north to around 2000 feet (612 metres) in the southern part. Geologically, the rocks of the Pennines, with some exceptions, tilt gently towards the east, but on the west they are cut off by abrupt downfolding or by faulting. The principal rock types comprising the hills are carboniferous limestone and millstone grit. Millstone grit (with some coal measures) is the dominant rock occupying the central section of the Pennines between the Craven district just north of Skipton and the more southerly Peak District. The shales and sandstones of this area form high hills occupied by moors and peat-mosses, with crags and exposed 'edges' of grey sandstone and stoney 'cloughs' descending to deep valleys. Over the years there has been the working of lead and other minerals on the hills, especially in the southern Derbyshire zone of the Pennines.

The research of Clarke, Roberts and colleagues suggest a 'phenomena zone' within the Pennines stretching broadly from the Derbyshire Peak District northwards to the Wensleydale area around Leyburn in North Yorkshire. Within this general region there are gaps, which seem rarely if ever to produce phenomena, and focal points of increased incidence of lights and possibly related or attendant phenomena. Purely for research practicality, Clarke and Roberts have broken down this general phenomena zone into six areas, some of which merge slightly with one another 'on the ground'. In the rest of this chapter we will work our way from south to north through these areas, looking at the testimony of folk-lore, place names and historical records, as well as modern case histories. For many years – almost certainly centuries – strange lights have appeared in the Pennines, becoming ingrained in the local psyche as surely as in the landscape. Some of the cases recorded here are among the most remarkable ever published, and yet the power of the testimony uncovered by the labours of Project Pennine lies not so much with these extraordinary reported incidents, as with the sheer volume of accounts for such a limited geographical area: it is that mass of testimony which leaves the reality of earth lights beyond doubts.

Map references are given for some locations, Plates 9–18 illustrate some of the places mentioned, and Figure 10 provides a general map of the region being studied.

AREA 1: EDALE (DERBYSHIRE) TO WESSENDEN MOOR (WEST YORKSHIRE)

This area primarily covers the western part of the High Peak of the Derbyshire Peak District, north of Buxton and east of Glossop, approx-

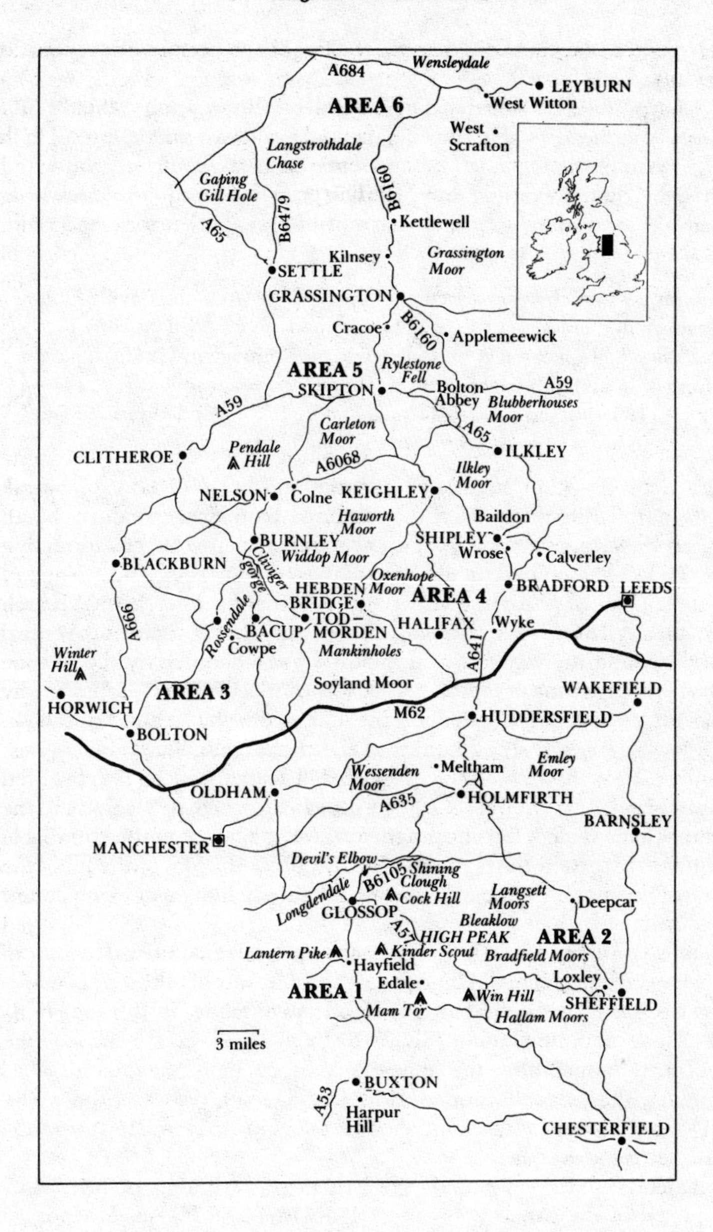

FIGURE 10 A simplified sketch map of the area of the Pennines found by Project Pennine to harbour zones of repeated earth light activity. Many of the locations referred to in this chapter are marked, together with some key towns and roads.

imately centred on the Longdendale valley (at the bottom of the map in Figure 10).

A sharp bend called the 'Devil's Elbow' on the B6105 Glossop–Woodhead road in the Longdendale valley, is said in legend to be the right arm of the Devil left behind when he had a confrontation with a weird light that prevented him chasing two lovers. John Davies, an 85-year-old local man, gave this account of an experience he had at the spot:

> I was on my motorbike on a section of the road known as the Devil's Elbow. The moon lit everything up as bright as day and as I rounded the corner . . . a great black wall appeared in front. I couldn't see through it. Had to stop right in front of it. It didn't frighten me but I had a queer sensation. It was like a massive black slug sliding across t'road and up t'moor. It disappeared and I got out and had a look. But there was nothing there.[1]

Kinder Scout, a high moorland expanse southeast of Glossop at SK 090880, derived its name from the Teutonic 'kunder' – creature, being, thing, prodigy. 'Scout' may be skuti or scoute – a cave or rock formed by jutting rocks. The prodigy in this case may be the 'mermaid' water nymph said to appear on Easter Eve at the Mermaid's Pool below Kinder Downfall, a gritstone rock waterfall (SK 083888). In the eighteenth century the area around the waterfall was believed to be haunted by the 'Kinder Boggart' (boggart means ghost, apparition in local dialect). Also on the southern face of Kinder Scout, on the 1500-foot (459 metres) contour of Grindslow Knoll, is a shack known as the 'Poltergeist Barn'.

Lantern Pike (SK 026882), a beacon hill to the west of Hayfield and opposite Kinder Scout, has a legend attached to it that 'Peggy with th' Lantern' (local dialect for ghost lights) swung its lamp on the summit on dark nights at certain times of the year. A pub at the foot of the hill. 'The Lantern Pike' has a signboard showing a figure holding a glowing lantern in his hand.

A lane running beneath the western face of Coombes Rocks (SK 018916), near the village of Charlesworth, southwest of Glossop, is named 'Boggards Lane'. Other potentially informative names in this part of the Peak include Shining Clough (SK 090980) which David Clarke feels was undoubtedly named after the appearance of the light phenomena which illuminates the whole mountainside (see below). Worm Stones (SK 043914), on Chunal Moor southeast of Glossop, may relate to the Old English term for dragon.

Longdendale is the valley of the river Etherow, a tributory of the Mersey. The stretch of the Etherow between Tintwistle and Pikenaze Moor has always been dominated by the frowning moorland heights of Black Hill and Bleaklow. Today's valley has the heavy, rumbling traffic of the A628 road, a necklace of five reservoirs, a railway line, and long stretches of high-tension electricity pylons. Odd happenings have been reported for centuries in this brooding, introverted place – ghosts of Roman soldiers, strange lights on the moors, and witchcraft. The 'Longdendale Lights', which have been appearing on Bleaklow for hundreds of years, are of several different kinds. One is a single powerful beam – rather like a searchlight – that has been seen in the vicinity of Bramah Edge and Clough Edge, beyond the Torside Reservoir. The other is a string of moving, elusive and eventually fading lights that often appear on the remote, craggy gritstone heights of Bleaklow beyond Shining Clough, in the area of the curious mound or hillock called Torside Castle. Rescue teams have turned out but no clue of any kind has ever been found which might explain the source of these lights.

Sightings of lights in relatively recent times for this part of Area 1 go back to at least the 1960s. In August 1962, for instance, then trainee teacher Michael Valance saw a 'box-shaped' object surrounded by light hovering above Kinder Scout. Phil Shaw of the Glossop Mountain Rescue Team (MRT) told Clarke: 'During the 1970s the National Park Warden (as they were then known) who lived at Crowden in Longdendale quite often saw lights (not flares) in the vicinity of Clough Edge above Torside reservoir. On occasions these were investigated, but nothing was found.' After the 1960s there has been a steady stream of sightings, and we give a brief selection here.

In July 1970 Mrs Barbara Drabble (a trainee teacher), wife of Peak Park Warden Ken Drabble, was driving towards Crowden Youth Hostel alongside the reservoirs on the Tintwistle to Woodhead road (A628) late at night:

> There was a good moon, and it was very warm. I had the car windows open. The light came from high up on Bleaklow and lit the whole valley for a couple of miles. I could have driven without headlights . . . then the air turned cold and I had to wind the windows up. It was terrifying and I didn't wait to see what it was.

In a telephone conversation with David Clarke in March 1987, Mrs Drabble told him that the 'light' referred to in the above newspaper account was not a single light, but that the whole mountainside was lit up just like daylight, stretching from Shining Clough across to Black Hill. When she passed the 'line' of the light on the road the air turned suddenly cold, her hair stood on end and she became terrified. When she emerged at the other side the car was covered in a thin layer of ice! 'Afterwards I asked local farmers if they had seen anything,' she said, 'but they wouldn't admit to seeing it and did not want to even discuss the matter.'

In July 1971 the same kind of light reappeared in the same place on Bleaklow, and this time was seen by several people at Crowden Youth Hostel. Mrs Joyce Buckley, the hostel warden, said: 'At first I thought it might be car headlights, but it reappeared on top of Bleaklow and no car can get up there.' A search party went out, led by Mrs Drabble's husband Ken, who said that, 'When we got to the top [of Bleaklow] there was nothing – no trace of people, lights or even a fire.'

This time several locals and farmers admitted to Mrs Drabble that they had seen the light on more than one occasion – but they wouldn't discuss it, 'their attitude was that it was something best left alone.' Farmers in the area are families who have lived in the valley for generations, and are very clannish and refuse to speak to outsiders and strangers about the 'lights'. Mrs Drabble did ascertain, however, that the light had been apparently appearing in the valley for hundreds of years, but had no specific periodicity in its appearances. There were also stories about sheep being found dead after the appearance of the light.

On 4 September 1972 a motorist reported to Glossop MRT that when he was driving on the B6105 towards Glossop alongside Torside Reservoir he saw a 'red flare' rise from above Clough Edge in the direction of Cock Hill. The area was searched without success.

On 21 July 1977 a former Glossop police sergeant saw what he described as a 'white flare' rising behind Shire Hill between Glossop Low and Hurst Moor on the western end of Bleaklow, just before 1 a.m. Another five independent witnesses in other parts of Glossop reported the same light. A search by Glossop MRT of the above areas could find no obvious explanation. At 5.30 a.m. that same morning, two textile workers in Hattersley, a village to the west of Glossop, saw a 'sausage-shaped' UFO disappearing into a bank of clouds.

David Frith, a member of Glossop MRT, first heard of the Longdendale Lights in 1972. Commenting in 1978 he said: 'There have been frequent calls about them to the rescue team. The last sighting I heard of was in October last year when they looked like a string of walkers carrying torches. They drifted about and then faded away. Other times it's been like a searchlight coming out of the hillside (Shining Clough).'

Glossop MRT was called out at 7.15 p.m. on 14 February 1982, after a sighting by Mrs Marion Lee of Torside Cottages, Longdendale. She saw what she described as a 'green flare' rising over the moors west of Torside Clough (above Bramah Edge). The search went on until around midnight, but no explanation for the light could be found. Bearings taken at the time from Mrs Lee's standpoint showed that the light had risen from the direction of Cock Hill. A Glossop MRT leader told a *Glossop Reporter* representative that: 'Over the years many stories have been told about mysterious lights on the south side of Longdendale and ghostly Roman legions in the vicinity of Torside Castle . . . I don't believe in ghosts, but there's something up there. It happens too often to be just chance.'

On 10 October 1982 Glossop MRT was alerted at 10 p.m. by a police officer, who when driving on the A57 Snake Pass towards Glossop over

Snake Summit, saw what he took to be a flare rising into the air between Janes's Thorn and the Lower Shelf Stones (a fault-line dissects the two rock outcrops). A search by the MRT north of the Snake Road failed to find an explanation.

Two doctors, husband and wife, were walking along the route of the Pennine Way over Bleaklow from Longdendale in April 1983. Having just passed by Torside Castle at 1 p.m. on a bright, clear sunny afternoon, they saw a silver-blue sphere moving towards them from the northwest, over Torside Clough, keeping close to the ground. 'It made no noise, and when more or less opposite us it stopped travelling horizontally and started to ascend. We stood and watched it ascend for about a minute or two. It became fainter and fainter, reflecting sunlight intermittently. It finally disappeared whilst apparently still ascending.'[2]

It is interesting to compare this experience with another a year later, about 25 miles (40 km) away on the opposite side of the moors below Kinder Scout at Edale. In the late morning of 27 August 1984, Mr Michael Davies, a Sheffield laboratory technician, was walking along the footpath from Jaggers Clough to Nether Booth in the Vale of Edale, Derbyshire, with a companion:

> At a point near Clough Farm (O/S map 110 SK 146865) I saw what appeared to be a spherical shining ball moving down the hillside from Nether Moor. This was approx. 200–300 yards away. The object moved over the top of the bracken. . . . I expected the object to get caught in the barbed wire fencing protruding above the top of the dry stone wall it was approaching. However, it did not, and proceeded downwards to the valley floor . . . .
>
> I now pointed out the object to my companion. We watched as it approached a clump of trees bordering the road from Hope to Edale. The object climbed above the trees, remained there for a short time before moving eastwards, apparently moving above the road. After a short time the object ascended rapidly upwards into the cloud and out of sight. The object did not shine continually, presumably it was reflecting sunlight intermittently.

Enquiries with Leeds Weather Centre ascertained that no weather balloons had been released anywhere near the area at that time, and meteorologists there were unable to offer any explanation for the sighting.

The northern part of Area 1, from Black Hill to Wessenden Moor, north of Longdendale, is one of the most remote landscapes in the Pennines. It is crossed by the A635 road between Holmfirth and Greenfield. There are several scattered reservoirs and pine plantations, but no houses at all except on the fringes of the towns at either end of the desolate tract. Geological survey maps show the area to be riddled with fault lines. A place name of interest is the 'Boggart Stones' (SE 024063), an outcrop of gritstone rocks on a fault, situated to the north of the A635 where it begins to descend into Saddleworth.

At Saddleworth, to the north, Running Hill Head Farm on the edge of the moors is haunted by a number of ghosts including 'curious lights' which were never very bright and moved about the bedroom like wisps of phosphorescent smoke.

A remarkable experience took place at a point on the A635 between SE 075072 and SE 051063 on the night of either 24 or 31 August, 1975. Service engineer Alan Fallows was returning home to Manchester, and after ascending into the Pennines out of the town of Holmfirth at about 8 p.m. he began to run into a thick hill mist. With visibility much reduced, he put his head out of the car window and crawled up the moorland road. He saw a bright light shining ahead of him through the mist. 'At first I thought it was the light of a farmhouse,' he said, '[but] when I got within about 200 yards I stopped the van. I didn't know whether to try to drive on and past or whether to turn around and go back. I was just so bewildered. It was massive, it must have been wider than the road itself, about 25 feet wide, 25 feet long and at least 15 feet high. It tapered towards the back like a pear or an egg. It looked as if it was fluorescent or transparent with a very bright light inside – like a light bulb. But it didn't hurt your eyes.'

The object seemed to be creating a 'tunnel' through the mist, and passed diagonally across the road in front of the witness' parked car, moving at a walking pace. It rose slightly above the ground as it passed above the wooden marker posts (about 3 feet – 1 metre – high) which mark the side of the road from the surrounding moor. The 'egg' then passed his left-hand side virtually upon the ground and, picking up speed, descended into the valley to the south of the road followed by two bleating sheep.

'I turned my engine off and then my headlights – I thought they might be attracting it in some way,' Fallows continued in his description. 'It passed about 20 yards in front of me and I saw a sort of "hole" underneath, a black circle about a foot wide, but there were no doors or windows. It took a minute to cross the road and I watched it for about four minutes altogether at close range. I was just terrified but it didn't seem to be interested in me at all.'

He found his ears popping, as if there had been a change in air pressure. Investigators also thought he may have had a period of missing time. Weather balloons were ruled out by Manchester UFO Research Association investigators – the witness had seen these before.

In March 1977 Miss Margaret Moore-Whittaker wrote to the *Manchester Evening News* to say that she had experienced a similar sighting, but Project Pennine have not been able to find reports of any other sightings in the area the night of the Fallows encounter. A policeman who used to patrol the A635 did tell them, however, that he had two strange sightings on that bleak moorland road – a bright light which zipped sideways over the bog for 300 yards (275 metres) and a circular light like a saucer-shaped spotlight which rose noiselessly into the sky at Wessenden Head (SE 075072). And

Park Ranger Ian Hurst told the Project team that MRTs twice turned out from Crowden in August or September 1975 to investigate 'flare-type' lights on the moors at Wessenden. They found nothing.

Jenny Randles, writing in her *Pennine UFO Mystery*,[3] notes that on three successive mornings – 15–17 January 1975 – between 3.20 a.m. and 5 a.m., strange white lights were seen moving over the moors between the Greater Manchester and West Yorkshire border. On at least one occasion a police patrol car gave pursuit, observing the lights to land on a hillside. Two policemen gave chase on foot across the dark and lonely slopes. But the light climbed off again into the inky blackness and was lost. This was interpreted at the time as one of the sightings of the so-called 'phantom helicopter' seen repeatedly in 1973–4.

The *Huddersfield Examiner* (19 September 1975) reported sightings of lights seen over the moors between Huddersfield and Oldham, and in August 1975 three multi-coloured lights were seen hovering at 200 feet (61 metres) above the surface of a reservoir north of the A635 (at SE 033083), reflecting on the water.

As a result of Project Pennine publicity, a woman telephoned Andy Roberts from Huddersfield to give brief details of many sightings by herself and members of her family of lights over the moors between the A635 and Meltham, to the north. Her brother apparently regularly goes up there to watch these strange lights on the moors. This is under investigation at the time of this writing, as is another case brought to Roberts' attention, in which a couple driving to Holmfirth on the A635 in October 1987 observed a very bright light, stationary in the sky. They stopped the car directly beneath it and could see lights above them in the shape of a circle.

AREA 2: DERBYSHIRE AND SOUTH YORKSHIRE MOORS

The region covered by this section of Project Pennine is slightly to the east of Area 1, stretching from Emley Moor on the West Yorkshire border, south into the gritstone Derbyshire moors to the west of Sheffield.

Many of the incidents reported here in recent times have been investigated by David Clarke and other active ufologists from the Sheffield and Chesterfield areas, and so are from first-hand sources.

Areas of traditional hauntings and folklore associations are so numerous for this area that they will be added in a later phase of the project.

We begin a brief selection of sightings for this area with a typical 'white lady ghost' type of phenomenon. The date was 31 January 1920, and the location was Long Lane, Loxley Edge (SK 306909). Farm worker Clarence Swain and his sister saw 'something white coming across like a woman maybe, then it vanished across by the old pit. It scared me right, and my sister couldn't talk for a bit as she was very feared.' Several other residents of the area also came forward to say that they had seen a similar ghostly white shape gliding around the old mine-workings on Loxley Edge (approx SK 308908).

In January 1985 two women on horseback on Loxley Edge were frightened by the appearance of a white figure which disappeared into thin air, causing the horses to shy. Several residents of Loxley village below the Edge have experienced all kinds of ghostly happenings and poltergeist manifestations. There was also a UFO encounter in the same area (see below).

At around 10.20 p.m. on 2 August 1958, a married couple saw an object between the Ladybower and Rivelin reservoirs on the Hallam Moors which 'spun like a burning light' across the sky towards Sheffield. 'I stopped the car,' said the husband, 'and we both got out to watch it spinning through the sky . . . . The top part of the object glowed like a neon light, while the bottom of it looked like a ball of orange fire.'

Mr G. Townsend of Mytham Bridge, Bamford, was standing outside his home around 8 p.m. on 3 March 1961, when he saw 'a large red object come from behind the end of Bamford Edge by Ladybower [reservoir]. It crossed the Dam and went behind Win Hill, then came into view again and continued up the Hope Valley till I lost sight of it. I called my wife out to see it . . . there was no noise.' Bamford Edge is a gritstone escarpment which has a legend of a 'grey lady' ghost, and Win Hill is reputedly haunted by 'Roman soldiers'.

In the early hours of 18 September 1973, a woman saw 'a phantom helicopter' rise out of a limestone quarry at Harpur Hill, just south of Buxton in Derbyshire. A security guard also saw something he 'assumed' was a helicopter as it 'seemed not to climb or bank as an aircraft would.' Reports were made to the police. The following month, police in the Buxton area received large numbers of reports describing a low-flying, hedge-hopping 'helicopter'. The police were unable to identify the craft, and witnesses described it as 'cigar-shaped' and around 40 feet (12 metres) in diameter, moving at high speed. The ghost 'helicopter' was still around by 14 January 1974, when police in the Macclesfield area of Cheshire received a midnight report describing an unknown helicopter, and were said to have 'kept the machine under observation for some time' as it manoeuvred over the Cat and Fiddle Moors, moving towards the Hope Valley. Derbyshire police were informed, and at 1 a.m. the crew of a patrol car spotted the 'phantom helicopter' flying above Mam Tor, in the Vale of Edale. The patrol car gave chase, but the object veered off heading towards Ringinglow, near Sheffield.

At 5.25 a.m. on 22 September 1975, a miner driving to work was near the TV transmitter mast on Emley Moor, southwest of Wakefield, when a

.ATE 1 The meteor of August 1783,
wed by some of the leading
entists and artists of the day from
northeast corner of the Terrace at
ndsor Castle. This aquatint is one
a series of illustrations of the pheno-
non carried out by Thomas
ndby, a founder of the Royal
ademy, and his brother Paul.
ndby, who personally witnessed the
nt, states at the bottom of the
ture that the object's apparent
meter was half that of the moon
t its light was 'much more vivid'.
e light emerged from a cloud. As it
gressed through the sky it 'grew
re oblong' and then 'divided and
med a long train of small luminous
lies each having a tail'. A full
ount is given by another witness in
apter 2. (*Reproduced by Courtesy of the
stees of the British Museum*)

.ATE 2 Crossroads on the A496
ween Harlech and Barmouth,
les. Balls of light were witnessed
ncing above the road here during
1904–5 light phenomena outbreak.
Chapter 3. (*P. Devereux*)

LATE 3 Egryn Chapel, alongside
e A496. This was a focus of light
enomena during the 1904–6
tbreak. Balls of light played around
roof, emerged from the fields on
her side, skipped along the hillside
hind it, and hovered over the road.
e Chapter 3. (*P. Devereux*)

PLATE 4 The chapel at Llanfair, south of Harlech. In 1905 balls of li[illegible] were seen to emerge from the field [illegible] the foreground, which is directly on the Mochras Fault. See Chapters 3 and 7. (*P. Devereux*)

PLATE 5 Three clergymen stood h[illegible] on Telford's aqueduct across the riv[illegible] Dee near Llangollen for several hou[illegible] one April night in 1905. They saw 'lurid' balls of light emerge from th[illegible] field below the aquaduct. Many of these flickered and went out, but so[illegible] gained in brilliance and flew off out [illegible] sight. See Chapters 3 and 7. (*P. Devereux*)

PLATE 6 The stone circle of Castlerigg near the Cumbrian town [illegible] Keswick. In 1919 a man reported a detailed sighting of extraordinary lights playing lazily about above thi[illegible] prehistoric site – see Chapter 3. (*P. Devereux*)

PLATE 7 Looking towards the Norman church of St Leonard's, Linley, through the gates of Linley Hall. This Shropshire hamlet was the scene of light and poltergeist phenomena in 1913. Balls of light were seen forming on the church tower and rolling down its sides. See Chapter 3. (*P. Devereux*)

PLATE 8 The church at Burton Dassett, Warwickshire, situated on a fault and a chief focus of light phenomena during the early 1920s. See Chapter 3. (*P. Devereux*)

PLATE 9 Carleton Moor in moonlight. This moorland, southwest of Skipton, has provided some of the most remarkable sightings - many made by groups of police officers - in all UFO literature. Some of these are published for the first time in book form in Chapter 4. (*Philip Mantle*)

PLATE 10 Looking west down part o
brooding Longdendale Valley, in
Derbyshire's High Peak. An ancient
focus of light phenomena, it records
the fact in place names and legends as
well as in modern sightings. See
Chapter 4. (*P. Devereux*)

PLATE 11 The remote A635 trans-Pennine road, cutting across
Wessenden Moor. Faults cross this
road, and many light events have
been seen out on the moorlands on
either side of it as well as above it. Se
Chapter 4. (*Andy Roberts*)

PLATE 12 The Boggart Stones,
Saddleworth Moor. 'Boggart' is a
local term for ghost, spirit, apparitior
This rock outcrop, on faulting, is
identified in legend as a place where
strange phenomena occurred, and
where they still do today. See Chapte
4. (*David Clarke*)

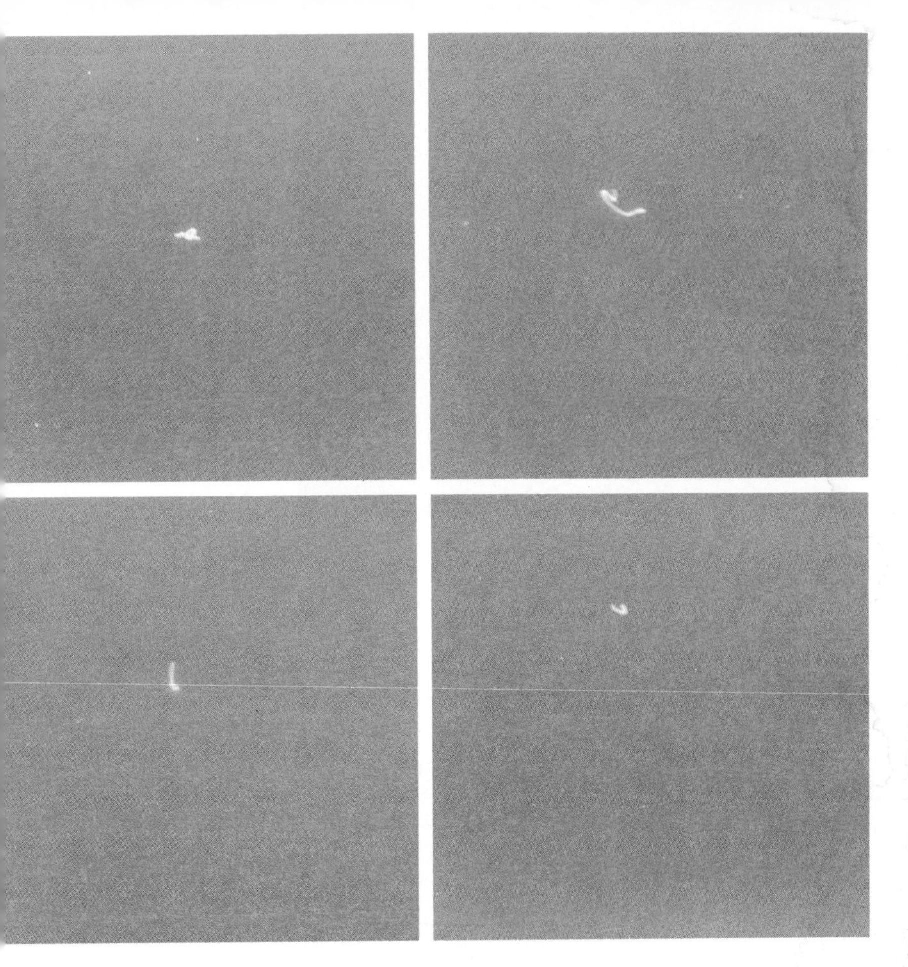

ATE 13 This sequence of photographs was taken of a light phenomenon witnessed
ar Rossendale, Lancashire. Account in Chapter 4. (*Copyright: David Milner*)

PLATE 14 Rylestone Fell, North Yorkshire. One of the several moorland locations around Grassington that
yields numerous light phenomena. (*Philip Mantle*)

PLATE 15 This remarkable and previously unpublished photograph of two balls of light flying over Carleton Moor near Skipton was taken by police sergeant Anthony Dodd. Note how the upper light illumines the underside of the low cloud cover. The picture was taken at 3.15 a.m. on 14 March 1983, with a Canon A1 50 mm standard lens, with a skylight U/V filter using Kodacolor 400 ASA film. (*Copyright: Anthony Dodd*)

PLATE 16 The Cow and Calf Rocks on the edge of earth-light-haunted Ilkley Moor. These rocks are a focus for light phenomena, particularly columnar effects. The outcrop also ha a recess known as 'The Fairy's Parlour'. See Chapter 4. (*P. Devereux*)

PLATE 17 The Devil's Elbow – a sharp bend on the B6105 in Longdendale. A fault crosses the road here, and strange phenomena have been reported at this point. The place has a legend associated with a magical light – see Chapter 4. (*P. Devereux*)

LATE 18 Emley Moor TV mast, uthwest of Wakefield, has been at ıe centre of several light phenomena 'ents. (*P. Devereux*)

LATE 19 The light tracks of :vorting 'Marfa lights' on Mitchell .ats, Texas. Light phenomena occur 'er a wide area of 'Big Bend' ›untry south of Marfa, and have ›ne since anyone kept records in the :gion (Chapter 5). In this photo-'aph, which was taken by Jim rocker over 3 minutes using a mm lens at f1.8 with 400 ASA film September 1986, distant car head-ghts can be seen as a straight line of ght in upper right. This allows com-ırison with the aerobatic trails of the ght phenomena.
(*'opyright: James Crocker*)

LATE 20 This light phenomenon as photographed by physicist David ubrin from the car park at Pinnacles ational Monument, close to the San ndreas Fault in California, in 1973 – ıis is its first book publication. The ght showed alternating effects of ass and weightlessness – see Chapter for a full account.
(*'opyright: David Kubrin*)

PLATE 21 A rare photograph of the 'spooklight' that moves along a road south of Joplin, Missouri, near Hornet. Dale Kaczmarek, who took this picture, has investigated numerous spooklights and debunked at least one. This one, which he observed at close quarters and throug binoculars, was the real thing. See Chapter 5. (*Copyright: Dale Kaczmarek*

PLATE 22 Physicist Harley Rutledg took this long exposure shot of a ligh moving in and out of clouds near Piedmont, Missouri, during a UFO 'flap'. We see the trail of the light he as it travelled towards a thunderstorn (distant light is caused by lightning). Rutledge and his team took many scores of pictures of lights during the 1973 outbreak of aerial phenomena i the region – see Chapter 5. (*Copyright: Harley D. Rutledge*)

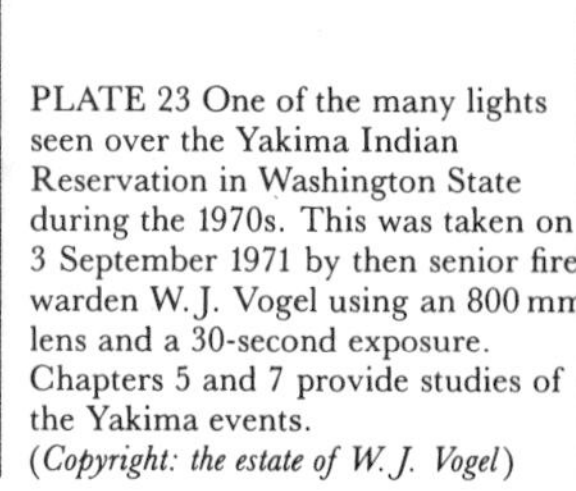

PLATE 23 One of the many lights seen over the Yakima Indian Reservation in Washington State during the 1970s. This was taken on 3 September 1971 by then senior fire warden W. J. Vogel using an 800 mm lens and a 30-second exposure. Chapters 5 and 7 provide studies of the Yakima events. (*Copyright: the estate of W. J. Vogel*)

PLATE 24 Pendle Hill, Lancashire, scene of geological faulting, spiritual visions, earthquake lights, elemental entities, witchcraft - and modern UFOs. A holy hill. See Chapter 6.
(*P. Devereux*)

PLATE 25 Some of Project Hessdalen's monitoring equipment inside their caravan HQ.
(*Copyright: Project Hessdalen*)

PLATE 26 Setting up powerful photography equipment at Hessdalen. (*Copyright: Project Hessdalen*)

PLATE 27 These pictures were taken by Leif Havik from Litlfjellet near Hessdalen in February 1984. In the first picture we see the light at the northern part of Aspåskjølen. It then moved towards Litlfjellet following the valley. The second picture was taken just before the light disappeared behind the flank of a mountain in eastern Litlfjellet. Havik used a Nikon camera with a 50 mm lens. Hessdalen is discussed in Chapters 6 and 7. (*Copyright: Leif Havik*)

PLATES 28 and 29 These two pictures (each with their close-up) were part of a sequence taken by Roar Wister on 18 February 1984 at Hessdalen. (*Copyright: Roar Wister*)

PLATE 30 The legendary Welsh mountain Cader Idris, its head in the clouds and its foot on the Bala Fault, viewed from the Barmouth area to its north. Paul Devereux and colleagues saw a ball of light erupt from the north side of this mountain in 1982 – see Chapter 6. (*P. Devereux*)

PLATE 31 The holy mount of Glastonbury Tor. A medieval earthquake shook down the summit church. Before the Christians the Tor was known as the palace of the Fairy King. Light phenomena of various kinds are seen every so often around this dramatic hill rising out of the Somerset levels. See Chapter 6. (*P. Devereux*)

PLATE 32 The production of tiny, short-lived lights in a laboratory rock-crusher in Denver, 1981. The streak coming down from the top of the picture is part of the fragmenting granite core. This is a frame from a slow-motion film made by Dr Brian Brady of the US Bureau of Mines. See Chapter 7. (*Copyright: Brian Brady*)

PLATE 33 In 1983 Paul Devereux, John Merron, Paul McCartney and colleagues also studied rock-crush lights. Various types of light effects were noted, some of which may have been artefacts caused by chips of incandescent metal from the rock-crusher's plates (see Chapter 7). Others were distinctly unusual light phenomena. This open-shutter picture by John Merron shows the trail of a point of light generated milliseconds before the granite core in the rock-crusher's chamber fragmented. We see the light swooped down to the bottom of the chamber, curved upwards and away from the camera. The secondary trails may be refractive effects in the rock dust or the plastic protective screen. The granite core, invisible in the darkness, occupies the centre of the frame.
(*Copyright: John Merron*)

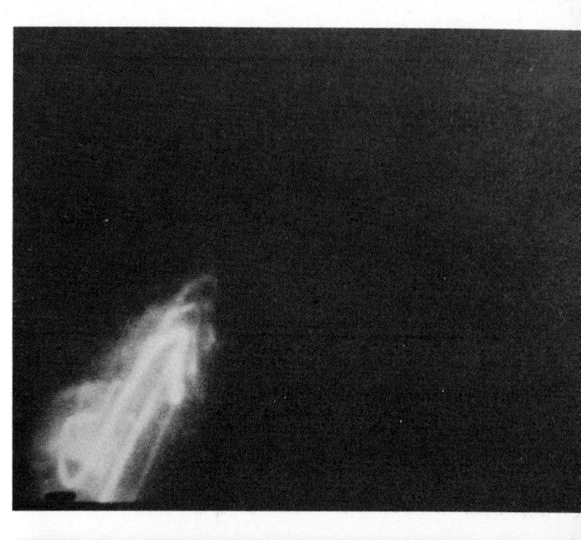

PLATE 34 A 6-inch long piece of rose quartz glowing after being subjected to friction (triboluminescence). Rocks do produce light! (*P. Devereux*)

PLATE 35 Milliseconds after friction, points of light gleam out amongst the dust above a piece of rock crystal, like bubbles in champagne. More tribo-luminescence – see Chapter 7.
(*P. Devereux*)

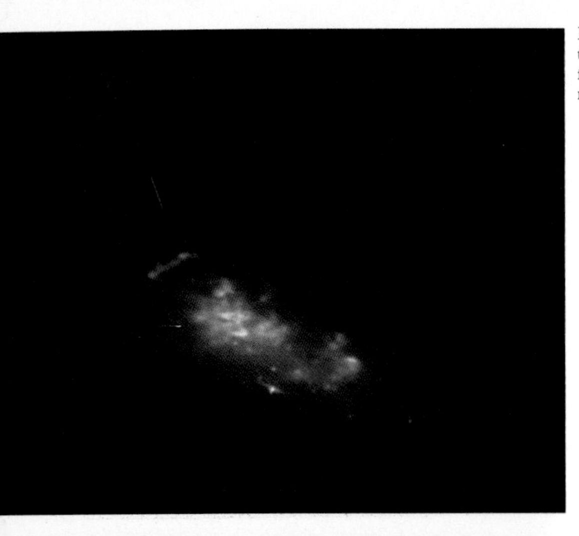

PLATE 36 Infra-red film cuts through the dust of a quartz block subjected to friction, revealing the points of light more clearly. (*P. Devereux*)

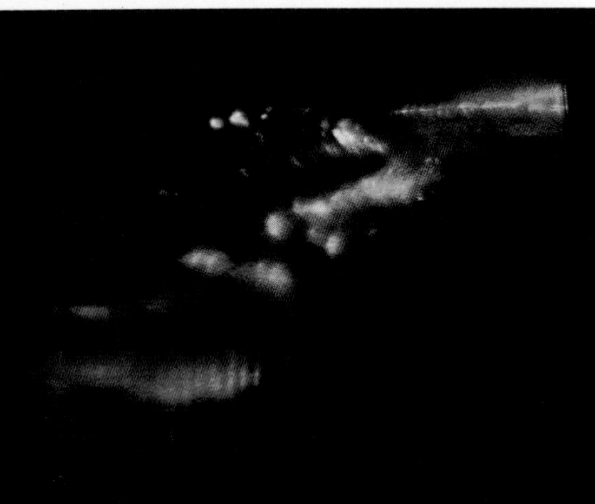

PLATE 37 Extreme close-up of surface of rock crystal moments after friction with another piece of quartz. (*P. Devereux*)

PLATE 38 Mount St Helens, displaying the 2000-feet-deep crater formed in its north side by the eruption of 18 May 1980. Located in the Cascade mountain range close to where Kenneth Arnold saw his famous 'flying saucers' of 1947, Mount St Helens reveals the active tectonic nature of the region. Before the 1980 explosion, researchers were getting reports of unusual lights and anomalous events. See Chapter 2. (*P. Devereux*)

PLATE 39 On the Yakima Indian reservation, Washington State, where earth lights were seen intensively during the 1970s (Chapters 5 and 7). One of the main foci of the lights was faulted Toppenish Ridge, part of which is seen here. The Yakima Indians vary in their feelings about the lights; some dismiss them as mirages, but others claim a traditional use of the lights was for divination: the colour, number and direction of travel of the lights were used much in the way the augurs of ancient Europe studied bird flight. Another Indian tribe of the American northwest, the Snohomish, consider the lights to be gateways to the Other World. The Lummis, on the other hand, have no extant traditions regarding the lights, though they are still seen occasionally above faulting off the coast of the Lummi reservation north of Seattle. (*P. Devereux*)

PLATE 40 Mount Shasta, northern California, said to be sacred to the Indians, and the southernmost major peak of the tectonically volatile Cascades range over which Arnold saw his flying discs in 1947. For at least a century there have been reports of curious lights clinging to the mountain slopes and the sudden appearance and disappearance of entities encountered by travellers. This sort of thing is typical of earth light zones, and is strikingly like the accounts relating to Pendle Hill, the holy peak of Lancashire. See Chapters 6 and 7 for peaks of illumination. (*P. Devereux*)

PLATE 41 The eastern end of the Chinati mountains in Texas, near the former mining community of Shafter. This area is a main focus for the so-called 'Marfa lights'. Minerals are a characteristic feature in earth lights geology. (*P. Devereux*)

PLATE 42 The notice board at the 'official' viewing point for the Marfa lights on Highway 90 between Marfa and Alpine. (*P. Devereux*)

PLATE 43 Looking across Mitchell Flat from the Marfa lights viewing point on Highway 90 between Marfa and Alpine, Texas. Lights are frequently seen skimming above the prairie. The Chinati mountains can be seen on the far horizon – they are a main focus for the lights. (*P. Devereux*)

brilliant bluish light materialised above it, about 1000 feet (305 metres) in the air, and then swooped down towards him lighting up the entire surroundings in such a way that he had to stop driving, temporarily blinded. Later he found a colleague had also observed the light.

The following month a couple driving on the A516 southwest of Buxton saw an object fly low over the road, almost hitting them. It had two rows of 'windows' separated by a bar, below which were flashing red, green and white lights. The faintly humming object moved rapidly through the night sky with its lights pulsating brightly, and seemed as big as a bus at close quarters. A geologist at Cowdale nearby saw the same object travel low across the moors, sweeping over him.

Loxley Edge displayed another of its phenomena in April 1977. A young couple parked in a car on Myers Lane around 9 p.m. heard a strange crackling noise and saw a 'bright orange light' appear a few hundred yards away from them over Loxley Edge. The 'large half-moon or dome-shaped object' moved towards their car. Silhouetted in front of the object was 'a large broad figure with a white haze surrounding its outline . . . which appeared to have frizzy hair and furry boots.' It appeared also to be over 10 feet in height! The terrified driver started up the car and sped along Myers Lane at full speed towards Bradfield, with the object following behind them, hugging the contours of the road as it moved. As they turned off the road at the end and looked back into Myers Lane, the object was nowhere to be seen.[4]

At 4 a.m. on a January morning in 1979, bus driver Mr Long, his conductress and a large number of passengers, parked at a stop at Wood Lane, Stannington, observed a number of strange lights manoeuvring above the valley of the river Loxley to the northwest.

In the summer of the same year Mr Billingham, walking along Loxley Road at 9 p.m., saw an object like 'a railway carriage brightly lit at the windows with a fuzzy outline' come forward slowly and silently in a shallow dive over Myers Grove School – the same area as the January sighting – on the far side of the river Loxley. It kept its 'nose' down and remained silently in this position for a minute or two, and them, suddenly, a red ball of light approached at high speed from the same direction as the object. As the red light neared the first object, it slowed down, passing over it, and entered the object. As soon as this happened the large object lifted its front and slowly it moved away, picking up speed as it climbed out of the valley and disappeared from view over the Bradfield moors to the northeast. Mr Simpson, a groundsman at Myers Grove School, has also seen strange objects in the valley when working early in the morning.

On the night of 12 January 1979, Mr AW, staying with friends in a scout hut at SK 240920 in Bradfield Dale, northwest of Sheffield, was outside with a friend when they saw a large disc-shaped object with a dome hovering about 100 yards (92 metres) away from them over the road to

Strines. It had revolving brilliant blue lights. They ran back to the hut to alert friends but the object had vanished when they looked again. The following night, Mr AW was riding alone on a motorbike on the road from Strines along the valley towards the scout hut. Near two reservoirs he saw, just 50 feet from him, a large metallic disc-shaped object covered 'with indentations like boxes and piping'. It was hovering and reflecting the moonlight, and passed out of view as he approached the hut. Terrified, he locked himself inside and subsequently experienced noises on the outside of the hut and the movement of the door latch. He fled the scene in the morning and has never returned.

Lights in the sky were seen over Burbage Moor, Derbyshire, in 1981, but police and Mountain Rescue Teams found no explanation for them. In September 1983, the Emley TV mast was once more the focus for a light display when a witness saw a huge blue glow in the sky close to the mast. There was no known meteorological explanation.[5] Again, in November 1983, a witness looking through his window towards the Emley TV mast at 7.13 p.m. saw a ball of brilliant orange light approaching it from a north-northeast direction at a very slow speed. The light passed behind the mast roughly half-way up the 865-foot (264-metre) tall structure and then headed towards the Holme Moss TV mast further in the distance towards Manchester. The man watched the object for three minutes through binoculars and saw no other lights that might identify it as an aircraft – just a single, perfectly round sphere of orange light. Further enquiries by investigators ruled out aircraft.

Another TV mast, this time at Unstone, near Chesterfield, became the location for a light phenomenon in the winter of 1983–4. Cutthorpe farmer Michael Aizlewood reported seeing a strange light darting across the sky. 'The pulsating light frequently changed direction and colour. It was at one time as intense as lightning and then it changed to orange. It seemed to be very close and once stopped as if to inspect the TV mast at Unstone. I couldn't make out any shape.'

An object 'like an upside-down ice-cream cone' seemed to be attracted to telegraph poles at Bole Hill, near quarries and reservoirs at Wingerworth, Chesterfield, in March 1983. A bus-load of people near Wingerworth reported a similar object. Chesterfield UFO investigators have looked into a large number of similar sightings in the area of the lead mines at Bole Hill.

At 10.15 p.m. on 9 September 1987 two witnesses independently reported observing a red light, which was at first thought to have been a parachute flare, in the sky over Big Moor, southwest of Sheffield. A search was carried out by Mountain Rescue Teams, but nothing was found either on Big Moor, White Edge, Barbrook or Stoke Flat. Froggatt Edge, part of the same escarpment, is said to be haunted by a White Lady ghost.

Two extraordinary incidents, reliably reported by multiple witnesses, occurred on the nights of 8 and 12 September, 1987 at Deepcar, near Stocksbridge. If true, and there seems no reason to doubt the veracity of the accounts, the events hint at a connection between what is termed 'fairy lore' and light phenomena.

On the first night, two security guards patrolling a section of the Stocksbridge by-pass road, under construction across the hillside north of Stocksbridge, saw a 'hooded figure' standing on a inaccessible bridge near the junction with Pearoyd Lane. Upon directing the headlights of their car upon it the figure vanished. Subsequently, both men saw what they took to be a group of small children dancing around an electricity pylon on the new road-bank. On approaching them they also vanished into thin air leaving no trace in the fresh mud. The terrified men reported the incidents to the police at Deepcar.

At midnight on 12 September, two policemen parked in a police car on the same stretch of road saw a 'shadow' circling around a white palate box below the bridge. It disappeared when headlights were turned on it. This happened three times. Shortly after this both men were terrified when the figure of a man dressed in white appeared by the side of their car; both saw it appear and disappear in a matter of seconds. It seemed to have a 'V' shape on its chest and was staring directly at them. This was accompanied by a loud impact on the back of the police vehicle. The policemen removed themselves from the scene.

At around the same time the above incidents occurred, two policemen reported seeing a number of mysterious lights moving around on a hillside at Lodge Moor, above the Rivelin valley in Sheffield to the south.

Four groups of witnesses in Wombwell and Hoyland, near Barnsley, saw a very large diamond or triangular-shaped object, black in colour and surrounded by a large number of brilliant multi-coloured lights moving just above rooftop height in a westerly direction on the evening of 10 February 1988. It followed the path of power lines towards Harley and Wentworth. Clarke has confirmed that no aircraft were in the area at the time.

The Wentworth/Kimberworth Park area to the southeast of this observation has had a number of 'orange ball of light' observations since the 1970s, including a great many in early 1988. One group of observers on 14 February, saw a large diamond-shaped, ruby-red object rise up from between two hills in the Wentworth area and then drop as if about to land.

AREA 3: LANCASHIRE

We now move northeast to the western foothills of the Pennines – mainly the Lancashire side of the moors stretching north of Manchester, through the Blackburn area, to Clitheroe and beyond. Reports from this area have been collected by many local UFO societies, most notably the Manchester UFO Research Association (MUFORA), who have investigated many of the reports around the Rossendale valley which became known as the 'Rossendale Anomaly', and featured in Jenny Randles's *The Pennine UFO Mystery*.

Four light phenomena foci emerge in this area:

1 The Rossendale valley, between Burnley and the area to the north of the city of Manchester. This is an area of high gritstone moors and beautiful valleys, heavily faulted and folded. Power lines, reservoirs and quarries stand out as areas where light phenomena have been reported frequently in the Rossendale area.

2 Rivington Moor. This high gritstone moorland height of Winter Hill above the towns of Horwich and Rivington, to the northwest of Bolton, is regarded as an area of consistently high UFO/light phenomena incidence.

3 The high gritstone moors stretching from Hartshead Pike and Uppermill, to the east of Oldham, to the northerly tracts above Littleborough towards Todmorden, reaching the border with West Yorkshire. All these areas have had a large number of UFO/light phenomena reports emanating from them every year since the late 1970s.

4 Pendle Hill and Forest. This area is probably the richest in folklore and historical traditions of paranormal phenomena in the whole of the Pennine range. It is dealt with in Chapter 6.

As with all the areas affected by light phenomena in the Pennines, accounts of what we would recognise as 'UFOs' go back decades – and the same phenomena given other names and couched in different perspectives can be found in the folklore and memory of generations.

The first date we pick here for a sample set of modern reports might as well be 20 July 1963, when a man at Cloughfold, Rossendale, saw what 'looked like a gigantic spinning top . . . the centre was of shimmering gold. There was a reddish glow at the top, and the bottom was a bluish-green. I know it was rotating. It seemed to hang in the sky – it was amazing. It was a clear night, and the object was to the east, towards Burnley, when I first saw it, but suddenly it arched across the sky toward Bacup, stopped, and shimmered again. It was very high and too big for a shooting star.' Local airbases could offer no explanation.

In May 1971 six Harwood residents watched an object which appeared over Winter Hill, near Horwich, just after midnight. Mrs Christine Campbell said that 'it hovered high above the television mast and glowed amber and made a buzzing noise . . . it was like a great lump of meat on a plate . . . it disappeared later but we stayed at the back of our house, waiting and watching for it to come back but it didn't.' Following this incident, several other residents came forward to say they saw a similar object a few days earlier. In August that same year, a group of witnesses saw a formation of four round glowing white objects in the sky flying directly towards Winter Hill.

The *Manchester Evening News* of 10 February 1979 reported a claimed phenomenon that might indicate unusual atmospheric electrical conditions. A Manchester woman and her two daughters had been terrified by glowing lights which hovered in Ashton cemetery, over the wall at the bottom of their garden. The tombstones had 'a bright orangey glow around them. It seems to come upwards from the stone, rather than above. It seems to be like a ray.' No logical explanation could be found.

In January that year two brothers were driving towards Burnley on the A671 near Water, Rossendale, when they saw a 'round glowing white object' 500 feet (152 metres) from the ground. 'It seemed to be hovering. As the road twisted at one point, the car headlights were pointing towards the object and it shot across the road.'

On 21 February 1979 at Stacksteads, Rossendale, tailor Mike Sacks and his wife saw at around 2 a.m. a pulsing light shine through the bedroom window. It was in a southerly direction, towards Lee Mill Quarry, on the hillside towards Littleborough. 'It was a brilliant pulsing white light with a slight orange tinge. It looked about the size of two double-decker buses. It was really weird. As we watched, the brilliant white light went out and a red undercarriage lit up dull red and started spinning round. As it spun round brighter red flashes were emitted.' In the space of about three seconds it descended vertically down, appearing to land in the quarry itself. At this point Sacks set out to fetch his photographer brother, and on their way back they met two policemen. The four men searched the quarry thoroughly without finding anything (though Mike Sacks had a strange experience – seeing a row of windows in what he thought was a portacabin, found not to exist the following morning). Nine other witnesses also reported seeing a bright white object in the sky, dropping down into the quarry. The quarry owners, Eskett – who extract gritstone from the hillside – had no explanations. Many other people throughout the west coast of Lancashire saw a brilliant roaring object in the sky the same night; in some cases the houses vibrated as it passed overhead.

Two young men on a fishing trip to the Ogden reservoirs, Rossendale, saw, around 9 p.m. on a cloudy, rainy night in August 1980, a 15-foot (4.5-metre) circle of red lights glowing beneath the surface of one of the reservoirs. The eight or nine lights were alternately 'dimming and brightening'. The lads fled, leaving their expensive fishing tackle by the side of the reservoir, not daring to return until daylight. Later that same night a woman at Haslingden, near to the reservoirs, was awoken by a red light flooding into her bedroom. Getting up and looking through the window she saw a circle of about 12 red lights.

A few days later Tim Edward, Michael Ashworth and Andrew Metcalfe saw an oval object flying over Newchurch, Rossendale, disappearing over the moors towards Cowpe. It had a bright light at each end and other lights flashing around its edges, and was between 50 and 100 feet (15–30 metres) in length. Ashworth remarked: 'It was definitely not an aircraft or a cloud. It seemed to be travelling fairly slowly and was quite low in the sky.'

Todmorden, some miles east of Rossendale, was the scene of a famous encounter between policeman Alan Godfrey and a brightly illuminated object in the early hours of 28 November 1980.[6] It had the shape of a spinning top, and was hovering about 5 feet (1.5 metres) above a deserted Todmorden road. Godfrey's first thought was that it was some kind of bus, but as he approached alone in his police car he was soon aware that he was confronting a completely bizarre sight. Both the policeman's radio-communication systems failed. He managed sketch the object. Godfrey was then suddenly aware that he was a hundred yards further down the road with only the dark night around him. This was later revealed to be a period of 'lost time' often experienced by close encounter witnesses, caused by amnesia. This missing period of memory was later probed by regression hypnosis and an abduction scenario ensued. (This aspect of UFO experiences is further discussed in Chapter 8.)

On 15 June 1981 a Rossendale woman saw a yellow, oval object moving against the wind in a westerly direction along the valley. It looked as if it were about to drop on the centre of Rawtenstall. Shortly after, in Rawtenstall itself, a woman saw a light shining through her window and looked out to see a very low-level yellow oval object surrounded by a mist or vapour. It moved over some trees above an artificial ski-slope on the hillside where it appeared to drop 'a spark' into the woods (probably those below Cribden Hill) and then vanish. Another witness nearby saw a 'golden star' approach the ski-slope, change shape to an oval with small lights underneath. It then seemed to drop a 'ladder' down into the wooded area on the hillside – like a streamer of light which unfolded and fell to the ground. The object then moved south and was seen by another witness at Helmshore.

On 9 March 1986 David Milner and four other witnesses saw a bright orange ball of light appear in the sky over Rossendale – 'the light kept coming on and then going off but it never moved. There is nothing but moorland in that area and we had no idea what it could be. I brought out my camera and took a load of photographs just using the 50-mm lens. I then took one photo using a telephoto lens and a 2-times converter. We must have watched this light going on and off for about 10–15 minutes before we got bored and went out as planned. We reported the sighting to the police and they informed us that they had a lot of similar reports, but we heard no more about it.' (See Plate 13). By early October 1987 light phenomena were still being reported in the Rossendale area.

AREA 4: WESSENDEN MOOR (WEST YORKSHIRE) TO ILKLEY MOOR (NORTH YORKSHIRE)

This zone is to the east of Area 3, stretching from the moors between Oldham and Huddersfield northwards past Bradford to the Ilkley district. It is an area so rich in anomalous light phenomena, perceived in different ways, that it is impossible within the context of this book to provide a meaningful list of even selected events. Instead, brief coverage is given to the main cases and areas. Full details of cases will be in the final Project Pennine catalogue.

Most sightings come from the moorland areas. Few are of 'structured craft', the majority being short-duration balls of light, of all colours. Nearly all of the areas mentioned below have had UFO sightings of one kind or another attributed to them since records began, and so folklore accounts are mixed in with the general area descriptions.

Stoodley Pike, a prominent peak above Todmorden at the head of the Calder valley, has been a monument of sorts for thousands of years. It lies at the edge of the Pennine escarpment. Folklore tells of strange flame-like lights seen playing around its base. Several 'ball of light' type UFO sightings have been reported from this area too, most recently to this writing in August 1987.

Widdop and Haworth moors, which lie above Hebden Bridge, have been the focus in the past of phenomena in which an appearance might be sudden, the light brilliant and fixed for a few moments and then, as suddenly, it would be extinguished, only to reappear some distance away. In the 1980s lights are still seen in this area and are interpreted as UFOs. One informant has indicated to Project Pennine that the Haworth Moor area and Hardcastle Crags are regularly host to blue balls of light.

The Cliviger Gorge, running from Todmorden to Burnley, has long been the haunt of all kinds of strange lights. A persistent and still-held belief is that the hounds of hell (Gabriel Hounds) sweep down the entire valley ending at Mankinholes below Stoodley Pike. Eagle Crag was said to be the home of one of the Pendle Witches, has a ghostly white form which flits around it, often in the shape of an animal, and latterly has been the focus for UFO sightings. Folklore populates the Cliviger Gorge with dozens of fairies, boggarts and Will-o'-the-Wisps. In the 1970s and 1980s it was the scene of many UFO sightings. The Alan Godfrey 'abduction' (included in the previous section) took place at the southern end of the gorge, and was immediately preceded by other witnesses seeing a bluish ball of light travelling down the gorge itself and others (police officers) seeing a blue ball of light travelling over Oxenhope Moor to the north which pulsated and zig-zagged across the sky, finally disappearing towards Todmorden.

Soyland Moor lies between Littleborough, Lancashire, and Todmorden. In the early 1970s, particularly 1974, it was a focus for what became known as the 'Phantom Helicopter' flap – strange lights seen throughout the Pennines, travelling at low level in the most appalling weather conditions. One notable sighting on Soyland Moor took place on 11 July 1980 when two policemen saw a rotating object low over some electricity pylons. The object changed colour from yellow to blue and then reversed the procedure, a typical type of sighting for this area.

Ilkley Moor is another of the moorland areas in West Yorkshire to attract UFOs. In particular the eastern escarpment overlooking the Aire valley where numerous UFOs have been reported over the years. The usual phenomena are small red or orange revolving balls of light. The Cow and Calf rocks, an outcrop overhanging the edge of the moor, have also had some intriguing light-related phenomena attributed to them over the years, particularly columns of light.

Although the earthquake-tectonic association with light phenomena will be more fully dealt with in Chapter 7, it is worth noting here that a flap of over a hundred UFO sightings in the Todmorden area, with various electrical malfunctions also being reported in association with some of them, culminated on 20 April 1981, coincident with a small earth tremor connected with slippage in the Craven Fault nearby. But the events surrounding 23 July 1984 are probably the best evidence of earthquake-related phenomena in the region.

The days previous to this period saw Britain's largest earthquake for over a century and was followed in succeeding days by lesser tremors (see Chapter 7). Local investigator Paul Bennett, editor of *Earth* magazine, made a special study of the short, intense flap between 21 and 24 July when hundreds of people across West Yorkshire saw unexplained lights moving about at low altitude. The focus of these lights appeared to be the south Wharf valley and in particular the area around Shipley. Paul Bennett has connected these with the local faulting (see Chapter 7). In one incident, around 10 p.m. on 23 July, Ian Tilleard and his girlfriend saw an intense white, small sphere of light, with a 'shimmering' edge, quite low down above Wrose, Shipley. They estimated its speed at about 30 mph (48 kph), and it was moving in a northeasterly direction. It travelled across moorlands into the darkness of Idle Moor and extinguished itself near Thackley Corner after about ten minutes. This was typical of sightings in the outbreak. A more dramatic case took place when a couple saw a bright light descend into Calverley Woods. It resolved itself into a 'UFO' from which entities appeared.[7] Calverley Woods, a gritstone outcrop, was the focus for a number of lightballs during the flap, and also has a long history of being haunted by 'The Calverley Ghost', which always appears along the same path in the woods (itself on a fault line) and is interpreted in the prevailing cultural terms. One local historian who collected accounts of the ghost said it was always described as a 'cloud-like apparition . . . simply a misty, impalpable shape'.

Judy Woods (or Wyke Woods) lie between Brighouse, Halifax and Bradford. The woods are about 2 miles long by $\frac{1}{2}$ mile wide (3.2 km by 0.8 km) and lie close to a reservoir and power lines. During the autumn of 1981 they were the scene of a small localised UFO flap. Bright streaks of light, similar to lightning, were the first light phenomena to be noted. Then witnesses saw lights flashing and hovering in the sky above the woods for weeks. 'Blobs' of light were appearing, splitting into smaller pieces, hovering and floating down into the woods. Some of the lights were interpreted as aeroplanes, others as UFOs. Over one hundred witnesses observed some part of the Judy Woods flap and photographs were taken which show streaks of light in the sky. Some sightings were accompanied by humming sounds and smells like 'rotten eggs' according to some accounts. This latter effect has been noted in earthquakes from time to time.[8]

AREA 5: CRAVEN HILLS - YORKSHIRE DALES

This region covers a large part of the Yorkshire Dales National Park stretching from Langstrothdale Chase southwards to the vicinity of the Craven Fault lines in the Skipton area. Grassington lies approximately in the centre of this section.

The geology is mainly limestone, although areas of gritstone rock occur near Harrogate, Blubberhouses Moor and Beamsley Beacon. It is a sparsely populated region, with a small number of policemen covering large tracts of farmland and moorland. The area appears to have the largest number of UFO and ball-of-light phenomena reports in the whole of the Pennine region. Several specific locations within this region seem to generate reports of these light phenomena year after year.

Craven fairies were said to be guardians of buried treasure, and were useful to miners in that they led to the richest lodes of metal. Both these traditions have been widely associated with light in many parts of the world. The fairies were said to live in caves and potholes, and the best time to see them was at midnight, as at cock-crow they melted into thin air. Places favoured by fairies were said to be Fountains Fell, Flasby Fell, the

Ribble Valley (Lancashire), Pendle Forest and the wild Pennine moors, including Trawden Forest, the heights of Cliviger and Rossendale – all of which were 'wick wi' fairies'. Coincidentally, these areas are exactly the same ones favoured by today's dancing lights known as UFOs. The late Francis King, known as the 'Craven Minstrel', firmly believed in fairies, and said he had seen them 'dancing' in the air at Lothersdale – a valley generating a large number of spooklight reports to this day. Kilnsey Crag, a limestone escarpment, was another site frequented by fairy lights, as was Rylestone Fell (also known as Cracoe Fell, between SD 990590 and 980570), a limestone fell topped with millstone grit rocks. According to local UFO investigators some people who live in Cracoe 'often refer to the dancing lights . . . .Legends dating back hundreds of years also mention strange lights performing in the night sky. There is no artificial lighting on the fell . . . the local populace have often noted coloured lights hanging on the face of the fell, and suddenly making ridiculous manoeuvres, gaining altitude and falling quite quickly.' A round green hill named Elbolthon, near Burnsall, one of a number of limestone reef-knolls along the line of the mid-Craven fault line, was said to have fairies inhabiting its limestone caves.

The famous Trollers Ghyll or Gill, near the village of Appletreewick and Skyreholme, is haunted by one of the most feared barguests or apparitions in the Yorkshire Dales; it is described as a large dog-like creature with round bright eyes as big as plates that apparently disappears in 'a sheet of flame'. Trollers Ghyll is a limestone gorge dissected by a fault. Place names in the area suggest that the legend dates back from the time of the Scandinavian settlers in the ninth century – hence 'troll', elemental spirit. The locality also has light phenomena events.

Almes Cliff (SE 268490) is a prominent group of millstone grit rocks, one of which is named the Altar Rock, and near to this is a natural opening in the cliff, about 18 inches (45 cm) wide, which is known as the entrance to the 'Fairy Parlour'. Grainge, the historian of Knaresborough Forest, said: 'The eyes of our forefathers saw fairies sporting over Almescliffe as abundantly as we see rabbits now.' (Cow and Calf Rocks, Ilkley – see Area 4 – also has a 'fairy parlour'.)

Light phenomena within the Yorkshire caves or 'potholes' have been reported occasionally in modern times, with people seeing things which would have been interpreted as 'fairies' in bygone days. Author Alex McClennan[9] saw a faint glow of light in Went Cave (SD 975745). It gained in intensity, was green in colour and seemed to be pulsating. The light grew stronger and moved towards him, at the same time a humming started which grew to a rumble. McClennan retreated hurriedly to the surface. Another incident happened in Gaping Gill (SD 751727), a limestone cavern on the southern flanks of Ingleborough. A potholer was last in the queue to be drawn up by the chair lift. He looked onto a boulder slope and saw a brilliant light, suggesting that another potholer was below ground. He went up the slope to where the light was shining, but it disappeared as he drew near. Using his own headlamp he saw the 'form of a monk' bending over into the boulder slope – it vanished in front of his eyes. Others claim to have seen ghosts in the same area.[10]

Even an abbreviated selection of individual cases of light phenomena encounters ought to start about a century ago. In the early autumn of 1897, for example, a farmer and his two brothers were salving sheep near the village of Burnsall, on the Craven Fault. It was a dark night, their only source of light being a lantern. Halfway through the job the light in the barn suddenly began to increase, and looking upwards they saw a brilliant sphere of dazzling brightness moving slowly above the barn. They described rays of light filling the entire building, and the brightness became so intense that it was brighter than a sunlit afternoon! The farmer said that all three of them were terrified. Slowly the light moved overhead in a straight line, a few feet above the barn, no noise was heard and eventually the sphere moved away into the distance. 'It was just a weird light that came and went,' said the farmer.[11]

Moving rapidly into the 'flying saucer' era, we have the case of John Kelly. In the spring of 1956, whilst sitting in his car near Burnsall, Kelly saw an object resembling two saucers, one on top of the other, flying over Simon's Seat (a prominent rock outcrop at SE 078598). The object hovered for a while and then flew off.[12]

A hiking party of four were returning to Ilkley from Bolton Abbey, Yorkshire, on 24 September 1956 when they saw a bright round orange sphere. It hovered for about five minutes before going below the ridge of a hill, 'very slowly and gracefully'.[13]

Two young mill workers saw an object hovering above Blubberhouses Moor on 11 December that same year. The two friends had gone for a motorcycle ride on the moor between Bolton Abbey and Addingham. 'Suddenly we were amazed to see a bright green object, which seemed to be hovering in the air about 2000 feet (600 metres) up. It was shaped like a ball and about 20 feet (6 metres) in width. Bright green and yellow flames were shooting from it.' The time was 8.15 p.m., yet the light from the object was intense enough to give the impression of daylight and the lamps on the motorcycle were switched off. The object hovered in the sky for about two minutes and then began to descend quite slowly. Whether it touched the ground or carried on its course over a fold in the hill, neither of the two witnesses could say. Corroboration for the sighting came from another witness who saw the object at precisely the same time as the two men, though from a much greater distance. The account appeared in *Flying Saucer Review* in January 1957, and even then the magazine could add: 'The Yorkshire Moors area contained between Grassington, Harrogate, Bradford and Skipton, seems very prone to saucer activity.'

For a set of more recent cases, a brief selection is made from the investigations of the West Yorkshire UFO Research Group (WYUFORG) and the Northern UFO Network (NUFON).

April 1970, Rombalds Moor, near Skipton: Mrs Wroe and husband were awaken from sleep by a 'high-pitched whining sound', and looked out of their east-facing window in Skipton to see 'a very bright ball of light' which hovered at a height of 'about 200 feet (60 metres) above the contour of the moor', at a distance of about 1 or 2 miles (1.5–3 km).

28 July 1981, Cracoe Fell: At about 2.30 a.m. two police officers travelling towards Grassington were passing Cracoe Fell to the east when a 'tower' of yellow-coloured lights appeared, hanging over the fell at low altitude. Six layers of lights could be seen on the object 'like a tower-block illuminated at night'. They watched this for 12 minutes until the object moved away at low speed.

March 1983, Grassington Hospital: On 27 March, nurses and assistants heard a number of strange siren-like noises coming from outside the hospital, and one saw a bright orange flashing light between a belt of trees outside. An examination of the trees later revealed a badly burned tree-trunk and several destroyed branches. On the nights of 30 and 31 March nursing assistant Paul Standage saw from the hospital, between midnight and 1 a.m., various flashing and flickering lights darting about the hillside between Rylestone Fell and Burnsall Fell (immediately to the south). They appeared to move in one direction and then the other. Later that morning another similar light was seen by two nurses over the moors behind the hospital which moved off along the valley to the south.

These were only some of the sightings observed in this immediate area over the early part of 1983.

April 1983, Low Cote Moor, Kettlewell: A farmer living near Kettlewell reported to local police sergeant Tony Dodd that one morning at 5.30 a.m. when leaving his farm to attend to his livestock in the fields he had seen a large glowing red sphere of light hovering in the air above a small stream which runs through his land. He said that the water beneath the sphere of light appeared to be hissing and bubbling as if it was boiling. The light then appeared to rotate on its axis and gave the appearance of being flat, and disappeared in the same manner as a flat object would as it rotated. The light performed strange movements, appearing to revolve, materialising and dematerialising as he watched it. After half an hour it disappeared.

1983, Yarnbury: Disused lead mines at Yarnbury (SD 009659) on isolated Grassington Moor have been the scene of repeated light phenomena. These moors are covered in old mine workings and are rich in various rare minerals. In 1983 a silent triangle of lights was observed above the old mines. Police officer Tony Dodd lists Yarnbury as an area that has a high number of sightings of light phenomena.

It was one of the places involved in an outbreak of light phenomena on 28

October 1983. Two stationary red lights and a large ball of white light were observed flying very low over the hillsides. The witnesses stated that the hills were illuminated. Thirty minutes later (6.30 p.m.) a witness reported seeing a glowing red ball of light near the mines at Yarnbury. At 6.45 two witnesses observed a glowing yellow ball of light over Cracoe Fell, followed five minutes later by another two witnesses reporting a yellow ball of light landing on Cracoe Fell. At 7 p.m. witnesses again saw a yellow glowing ball of light hovering at Yarnbury. The following night at midnight witnesses on Carleton Moor saw a glowing ball of yellow light which appeared to hit the moor and then move off again. At 12.30 a.m., just north of Skipton, a police officer reported seeing a glowing triangle of white light, very large in size.

29 October 1983, Summerscale, A59 Blubberhouses Moor road: At 4.38 a.m. on a frosty night with a clear sky, two police officers on routine patrol were startled by a large glowing object hovering in the sky, moving behind gaps and folds in the hills in the distance. They described the object as resembling a chandelier, with multi-coloured lights flashing on and off; the lights were like two rows of peas one on top of the other, the top row being slightly shorter than the bottom row, each light flashing a different colour in an irregular pattern. The object was apparently about 5 miles (8 km) away from their position and very large in size. Although a search of the area was made and enquiries made with military and civilian airports no explanation could be found for the sighting.

7 November 1983, the B6160 Addingham Road, between Grassington and Ilkley: Police Sergeant Tony Dodd and his wife, Pauline, had left their home in Grassington and were travelling to a previously arranged meeting in the small town of Ilkley, through an area from which unusual light phenomena had been reported before. Dodd is an experienced photographer and always carries a loaded camera.

As the vehicle passed near Bolton Abbey ruins, Mrs Dodd noticed a cluster of brilliant red-coloured lights moving slowly between the darkened hills to the west. To the naked eye the object appeared to be round. The silent lights passed about a mile away from the witnesses, who stopped the car to watch them through binoculars. With magnification, there seemed to be hundreds of red-coloured lights which were all pulsating in unison on a 'top-shaped' object. The lights appeared to descend to around 500 feet (150 metres) between Beamsley Beacon and Beacon Hill.

After continuing the journey for another mile, the lights suddenly came into view again. At this point the car was stopped and three photos were taken (Canon AE1, 1000 ASA film, 50 mm lens at f2.8, shutter speed 1/16th second, hand held), before the cluster of red lights disappeared behind the moorland north of Addingham. The lights moved slowly with no sound at all. Later checks made at military and civil airports ruled out any form of aircraft. Nothing had been picked up on radar.

Computer analysis of the three photographs was conducted by the American group 'Ground Saucer Watch'. They concluded that they 'represent Britain's first UFO photos'. In summary, breakdown of the image on computer revealed the major part of the object to be 'tenuous, cloud-like in structure', a rounded image revealed by analysis was detached from the 'vapour' and appeared to be the only 'solid' part of it . . . 'there is no evidence of a top-shaped object – only a round object at best'. The vapour, when matched against several different kinds of known gases, did not correspond to any.

29 February 1984, the B6265 Skipton-Grassington Road: Two police officers parked near the village of Cracoe in the early morning hours saw a large white glowing light in the sky over Bordley Moor to the northwest (SE 964605). As a plane came into view, a smaller ball of white light emerged from the larger one, then both moved off in opposite directions, the larger going south towards Skipton.

Finally for Area 5, we come to Carleton Moor. This region has been left until last and treated separately, due to the large number of reports of low-level luminous phenomena reported from it. Virtually all of the incidents recorded from this area have been investigated by Tony Dodd, who is based at Grassington. He and other local investigators, as well as other police officers, have seen a large number of light phenomena in the area for themselves.

'Carleton Moor' is actually a name used to describe a larger area of moorland, about 5 miles (8 km) in length, stretching southwest from Skipton towards the north Lancashire towns of Earby and Barnoldswick. The moors reach their highest at Pinhaw Beacon (1273 feet), a modest height commanding a wide and varied view. The moors of Carleton, Elslack, Cononley and Thornton have figured prominently as areas where mystery lights have been reported by local residents in the late 1970s and early 1980s. During the 1983–4 period Yorkshire UFO Society (YUFOS) investigators established that unusual light balls have been seen by locals for generations. One old gentleman cycled up to the skywatch caravan to inform the investigators that he had seen unusual objects in the area well before the Second World War, and in 1984 Philip Mantle and other YUFOS members spoke with local CB radio enthusiasts who use the moortops for conversations due to the good reception there. When asked about the local UFO sightings they replied, 'Oh, you mean the flying oranges?', in reference to the small orange balls of light they regularly saw manoeuvering above the moortops.

A common area for reports of light phenomena is the Lothersdale valley, below Carleton Moor, which is scarred by many disused quarries. Policemen and others have observed large and brilliant 'chandelier'-like objects above the quarries in the valley, between Raygill and Lothersdale.

Tony Dodd says in reference to the Carleton Moor area:

I have personally witnessed these balls of light many dozens of times, always with witnesses, and have over the years given a great deal of time to their study. I have seen them at close range and . . . have photographed them.

At the present time I have come to no logical conclusion as to the composition or origin of the phenomenon but I feel certain that we are dealing with an intelligence which is totally outside the scope of our understanding or science. I do not make these remarks easily . . . .

From a geographical point of view the areas of observation contain open moorland, plantations of fir trees, reservoirs and in some cases ancient burial grounds. The location of the light balls is usually the highest in the area of Carleton Moor, near the large microwave repeater station aerial (SD 943462), along the Wharfedale Valley towards Grassington. No sound is usually heard when the objects are observed, and the majority of them seem to be travelling 400 feet above the ground. Many of the witnesses state that the lights seem to be following the contours of the moors.

I have found over the years that [the local farming] communities are very superstitious by nature and tend to avoid talking about events for which they can offer no logical explanation.

Space precludes an adequate summary of Carleton Moor sightings, which are amongst the most remarkable in all UFO literature, and the following are merely to give a flavour of the kind of phenomena reported.

January 1978, Cononley and Carleton Moors, near Skipton: Police sergeant Tony Dodd and PC Alan Dale were driving along an isolated country lane outside the village of Cononley when suddenly the dark road in front of them became lit up with an unusual light. The weather was fine, dry and very cold. They then saw an object appear in front of them about 100 feet (30 metres) away, moving at under 40 mph (64 kph). It had 'a bright incandescent glow' and it passed over the police officers' heads. There were 'three great spheres underneath, like huge ball-bearings – three of them equally placed around it.' The object appeared to have portholes around the dome and coloured lights – blue, red, green and white like neon lights which gave the impression that the object was revolving. It was completely soundless. The object moved over the moors to the west and appeared to 'go down' into the Standrise Plantation on Elslack Moor on a distant hillside. Afterwards they met another police car whose occupants had seen the same object landing in the same area.

March 1980, Standrise Plantation, Elslack Moor: Mr X and his family were travelling along the road that leaves the village of Carleton near Skipton. The road rises up and over the moors near Elslack. The family were heading for the supermarket in Colne – a regular journey. They had neared the Standrise Plantation when Mr X looked through his mirror and noticed two coloured balls (red and green) moving closer to the car. The balls then moved level with the car. They were the size of footballs, one situated on each side of the car. They continued to fly alongside the car at window height; the occupants became very frightened, and Mr X found it very hard to keep the vehicle under control. Mrs X was shouting at her husband to turn the car round and return home, the two children were screaming with fright and the dog cowering. When eventually the car turned around, the two balls of light continued to follow it. They tracked the vehicle for a short while before disappearing.

10 March 1981, Carleton Moor: Two police officers observed a large glowing oval shape moving slowly over the moor in a north–south direction. Attempts were made to follow the glowing sphere but these were in vain. The light moved as if under intelligent control, appeared to be around 12 feet (3.6 metres) in diameter with no apparent structure, moving leisurely about 500 feet (150 metres) above the moortops. Enquiries ruled out aircraft of any kind as an explanation.

August 1981, Carleton Moor: A couple were out walking their dog on the northwestern side of Carleton Moor on a dark night with very clear sky. They noticed two lights on a hillside going on and off alternatively at approximately one-second intervals. They then saw another identical light. The lights were spread out in a triangular fashion, each light being approximately half a mile distant from another.

By the time the witnesses reached the edge of this triangle, five balls of red light had appeared within it. The balls of light now moved closer to the witnesses doing aerobatics all the time. The cows and sheep were now acting disturbed, the couple's dog was barking, and they began to back away. Suddenly white beams of light started coming from the sky at all angles, hitting the ground at the far end of the triangle. The beams remained visible for some time. They were described as narrow at the top and wide at the bottom. They appeared to make a slight 'zipping' noise. The flashes moved within 100 yards (90 metres) of the frightened witnesses and they ran to a nearby church (SD 893480) to continue watching the spectacle. They then became aware of a large cloud moving into the centre of the triangle, its underside glowing red. The red balls of light close to the ground took up positions under the cloud, and moved into it. The flashes stopped, the pulsating white lights on the ground had gone and the cloud moved away at fast speed. All the animals settled down. Tony Dodd, who personally conducted the investigation into this case, vouches for the witnesses' credibility.

4 March 1982, Coniston Cutting: Mrs Y, a business-woman from Skipton, was returning home from Lancashire on the A65 road, deserted at about 10.30 p.m., and was between the villages of Hellifield and Coniston Cold, passing through 'Coniston Cutting' where the road cuts through the hillside. The car was lit up with a bluish light. Looking in the mirror the woman saw two lights – blue and red in colour – hovering in the air behind the car about 1 foot (30 cm) apart. After following the car for about half a

mile, a beam of light suddenly appeared on the nearside of the car, narrow at the top and wide at the bottom, forming a circle of light on the ground 15 feet (4.5 metres) in diameter, lighting up the surrounding fields. This light was coming from above the car.

Mrs Y then began to fell very cold, and the engine of the car began revving as if it had been lifted off the ground. The lights disappeared as a lorry approached in the opposite direction, but upon returning home the witness found that she had lost 30 minutes of time, and subsequently suffered a series of physiological symptoms including a skin rash and energy loss. In May 1982 she was subjected to regression hypnosis and questioned by YUFOS investigators.

Under hypnosis the witness described seeing the lights above the car, experiencing strange physical sensations, including a feeling of intense pressure between the eyes. Although she had no recollection of being aboard a craft or seeing 'creatures' she began to speak as a channel between the questioners and the 'abductors', including one entity calling himself 'Zeus' who claimed to be from the 'planet Zircon', as well as several other mythological names.

24 October 1982, Elslack and Carleton Moors: In the early hours of the morning, three police officers parked in a patrol car on White Hill Lane observed a small object zig-zagging a few feet above the ground across Carleton Moor (at approximate map reference SD 965475). It appeared to flash a beam of light down towards the ground at intervals, as if the light was searching for something. Red and white lights could be seen revolving around the body of the object, which was emitting a beam of light from its underside directly onto the ground. At one point the lights came within 50 feet (15 metres) of the witnesses.

At 5.15 a.m. at map reference SD 965475 a series of stange lights passed directly over the police car. They were not seen approaching, nor was any sound heard, although it was very low. The men elected to try and follow the lights, and drove down a steep hill (Babyhouse Lane towards Lothersdale) for a few hundred yards before losing sight of it. Five minutes later, the men looked back to what had been their position on the hillside. They could clearly see a single 'strip' of light, crescent-shaped, travelling slowly towards them. The bright strip of light did not give off a beam, but it came directly over their car, some 75–100 feet (23–30 metres) in height. One man commented, 'It was so low you could have thrown a rock at it.' The light now became much brighter, as it passed over the bemused and somewhat frightened men. An array of red and white lights were clearly visible on the underside perimeter of the object.

14 March 1983, Carleton Moor: While on patrol duty, Tony Dodd saw two lights pass over the moor at 3.15 a.m. One of the lights illuminated the underside of the cloud base. Dodd was able to take a photograph using a Canon A1 with a 50 mm standard lens and Skylight ultraviolet filter. Film

was Kodacolour 400 ASA. This picture is published for the first time in Plate 15. Dodd, like all those who have allowed their pictures to be used for this book, asked no fee.

11 August 1984, Carleton Moor: At 1.45 a.m. Mr and Mrs W awoke to see from their bedroom in Carleton village a bright light emanating from the Standrise Plantation to the southwest. As they watched they saw a series of three large sphere-shaped lights hovering above the trees in the plantation. They moved towards the village and the couple then saw a large disc-shaped object behind them. It seemed to them to be about 100 feet (30 metres) in length. The centre light pulsed, followed by three flashes, then went out, followed by the outer lights pulsating. After about a minute the object began to move erratically over a small area of land and then flew away to the west without making a sound.

17 May 1987, Earby Moor: Mrs Jean M, living in a cottage on the edge of the moors, was awakened between 2 and 2.30 a.m. by a loud crashing noise and found that a chopping board had fallen off a cupboard and smashed a glass. Shortly after this she heard a loud droning noise (unlike an aeroplane engine) passing over the cottage and looked out of the window to see a very brilliant green light in the sky: 'it wasn't like a solid object.' It flew towards Kelbrook.

AREA 6: WENSLEYDALE, NORTH YORKSHIRE

This is the northernmost locality for repeated earth lights occurrence noted by Project Pennine. This isolated region is approximately centred on Leyburn, in Wensleydale. It has not yet yielded the large files of individual cases in the manner of the other Pennine regions, but certain places have long been noted as haunts of light phenomena. Writing in 1968, for instance, J. Harris noted:

> The A684 . . . runs from Northallerton right across the Pennines to Kendal. The stretch between Leyburn and Swinithwaite, notably on each side of the village of West Witton, is illuminated paranormally on some winters' nights. Motorists . . . have seen a source of light which is on the road and is so bright that they have flashed their headlights, believing that an oncoming motorist has forgotten to dip his lights. As they slow down they realise that the light is not moving. The light always vanishes abruptly.[14]

In the 1950s, A.A. MacGregor noted another such light a few miles away near West Scrafton, Coverdale:

> I have before me the written and signed testimony of Mr. William Brown, a

> contractor at West Scrafton. 'I have seen the Pennine Light near Scrafton, usually at Christmas time,' he writes. 'A brilliant light something like that of a car. As one gets nearer it vanishes. It is rather difficult to explain.[15]

McGregor noted that the 'Pennine Light' was 'well-known'. W. R. Mitchell also refers to a ghost light in Coverdale.[16]

5 AMERICAN SPOOKLIGHTS

The USA is another country that has a reasonably accessible catalogue of earth lights locations: there are said to be over a hundred such places known there, though doubtless this is only a small percentage of those which actually exist. The lights are usually referred to as 'spooklights' or 'ghost lights' in America, and the frame of reference is nearly always supernatural. They are seldom associated with UFOs, which, in the USA possibly more than anywhere else, are almost always assumed to be extraterrestrial.

These American earth lights vary like everywhere else in the world between appearing fairly regularly in one small area over a long period, to occurring intensively in a region for a specific period of weeks or months. And, of course, there are always the mavericks – the isolated instances of light phenomena. In our selection below, we will look at all these kinds of earth lights occurrences in America.

TEXAS

Possibly the most famous American spooklight location occurs in the 'Lone Star State', around Marfa, in Presidio County about 200 miles (322 km) southeast of El Paso near the 'Big Bend' area of southwest Texas. A popular observing point is on Highway 90 about 7 miles east of Marfa towards Alpine. A notice by an abandoned airfield has been erected indicating the sighting point for the Marfa lights. They usually, though not always, appear to the southwest in the direction of the Chinati mountains, about 50 miles distant across Mitchell Flat. They do not appear every night, but are said to be quite regular. A great many people, both locals and visitors, claim to have seen them.

The first recorded report of them seems to be that of Robert Ellison in 1883, who at first thought he was seeing Apache campfires, but soon realised they were something far more mysterious. He saw them as glowing balls at the base of the Chinati mountains, and said they floated up and down and bounced back and forth.[1] Other settlers before him in the region said they had always seen the phenomena. The lights usually have a yellowish cast, and flicker and frolic around the desert. 'It looks like a big headlight,' says Hallie Stillwell, who has lived in the country for over 60 years. 'It just kind of flickers along the mountain. . . . They light up and run across the mountain, kind of like a grass fire.'[2]

A filling station owner from Alpine, driving home from El Paso with his wife on Highway 90 west of Marfa, saw lights coming up behind him through his rearview mirror. Thinking it was a truck coming dangerously close, he peered over his shoulder only to have the lights suddenly disappear.[3] 'Those lights have made good Christians out of a lot of people who weren't before,' the garage man noted wryly – a similar effect that the Egryn lights had on the people around Barmouth in Wales!

As usual with American spooklights, the sceptical explanation is that the lights are car headlights refracted by air layers of different temperatures. In this case the culprit vehicles are supposed to be running along Highway 67 linking Marfa with Presidio to the south. It is true that this road cuts across the line of vision towards the Chinati mountains, but it will not do as an explanation. To begin with, there were no cars, or electrical lighting, in the area when the lights were first reported. In addition, people have had close encounters with the lights. As one witness said; 'When you see them, you know you are seeing them. There's no doubt.'[4]

One of the best instances of close observation came as a result of the investigations of two geologists, Pat Kenney and Elwood Wright, who were carrying out geological work in the area in the 1970s. Hearing about the lights, the pair carried out a number of observations between March and June 1973. They went, sometimes with companions, to the viewing point on the Alpine road on several occasions during March, and did indeed see flickering lights in the distance. But they simply could not be sure they were not car lights, and felt inclined to dismiss all the lights as such, even though they knew that there were no roads in the areas where they saw some of the lights. But on 19 March 20 minutes of viewing was sufficient to convince them that this was not the answer: the lights began swinging in an arc 'like a rocker on a rocking chair', and one did a complete loop. 'They appeared to be playing,' the geologists commented. The lights were still being observed at a considerable distance, though, so the following evening Wright and Kenney decided to try to get a closer look. At least three lights were visible again in the Chinati direction. They drove along a dirt road a few miles onto Mitchell Flat. As it was bright moonlight, they drove without headlights. They stopped, and waited. Three range horses near their vehicle suddenly acted in a startled manner. Wright and Kenney, writing in the third person, take up the account:

> At this precise moment, they observed two lights moving rapidly from the

> southwest to the northeast, almost at right angles to the road. The first light slowed down near the road, crossed the road less than 1000 feet in front of their car, and continued to the east where it seemed to merge with or meet a third light which was brighter and was between their car and the vicinity of the old Air Base hangar. It crossed the road only three or four feet off the ground. The second light coming from the southwest followed approximately the same path as the first light but it seemed to be moving slower. They decided to try and sneak up on it in the car with the lights out, and try to intercept it at the same place where the first light had crossed the road, but they could not get close to it . . . it started veering to the north and when they got to the point where the first light had crossed the road, the second light was several hundred feet down the road. They could tell how far away it was because it was only about three feet off the ground and it went behind some bushes and in front of other bushes . . . . The second light was not more than 200 feet from them when it crossed the road. They stopped the car and turned off the engine as it came to the edge of the road. It moved out to the middle of the road and hovered there. They both had the distinct impression that it knew exactly where they were and that it was just daring them to chase it. The intensity of the light decreased as it slowed down and hovered in one spot. The color was approximately the same as ordinary incandescent household light bulb. It seemed to possess intelligence! They did not move and after approximately one-half minute it moved on to the east to join the other lights, then they all vanished. The light they observed closest appeared to be about half the size of a basketball.[5]

The observation of apparent 'intelligence' on the part of the lights is worthy of note. Kinney expanded to a news reporter that, 'it had intelligence, definitely . . . ', and Wright said that, 'I really and truly don't have any idea what it was. It kind of looked like it was playing with us. It was a heck of a lot smarter than we were.'[6] UFO researcher and journalist Dennis Stacy likewise makes reference to this characteristic: 'the lights also appear to have a crude capacity to interact with their human observers, either moving farther away as one approaches, or disappearing.'[7]

A group of Marfa lights were photographed in 1986 by James Crocker (Plate 19). His long exposure picture not only shows the aerobatic nature of the lights' movements, but allows these to be compared to car headlights in the distance. In distant views, such photography can show the difference between headlights and the anomalous lights more surely than the human eye.

The popular context of the lights is, again, that of ghostly manifestations. But the Marfa lights have generated a particularly rich array of legends, both old and modern. It is said that the lights were known by the Mescalero Apaches of the region, and were perceived as spirits by them. The lights were claimed by settlers to be, variously, the ghosts of massacred Indians, massacred Whites, the ghost campfires of a destroyed wagon train, guardian lights of hidden treasure (as in the Old World belief systems), and much else besides.

Other Texan spooklights include 'Bailey's Light', a basketball-sized glowing orb sometimes seen near Highway 35, 5 miles (8 km) west of Angleton, south of Houston and close to the Gulf of Mexico shoreline. This is supposed to be the ghost of a nineteenth-century settler called Brit Bailey, searching for a jug of whiskey! Not far away – in Texan terms at least – is the Bragg Road Light. This is seen on an 8-mile long (13-km), dead straight road that links Highway 787 with 1293 near Saratoga, northwest of Beaumont. The sand road runs along the course of what was formerly a rail track, in the heart of the dense forest known as Big Thicket. People for decades have seen a strange light moving down this road. It has been seen as dull yellow, white and red. In 1973 a photographer taking pictures for *National Geographic* saw it, and reasoned it had to be 'diffused' light from distant car headlights. But those who claim to have had close encounters profoundly disagree. One car load of witnesses saw a bright light ahead of them on the road at about 'head height'. As they drove towards it, it moved ahead of them and then went out. Suddenly, and inexplicably, it reappeared behind them.[8] The main legend associated with it says that the light is the ghost of a brakeman who worked on the former rail line. He was decapitated in an accident, and his spectre now wanders with a lantern looking for his head. Another folktale says that it is the ghost of a Spaniard guarding treasure. Apart from refracted headlights, the other sceptical explanation is that the light is a form of Will-o'-the-Wisp, as there are numerous swamps within Big Thicket. However, the same explanation is trotted out for the Marfa lights as well, where marsh conditions simply do not exist.

Other spooklight locations in the state are claimed at Esperanza Creek in La Salle County, southwest of San Antonio, and at Sarasota.

COLORADO

Blue-tinged balls of light were first noticed in the cemetery at Silver Cliff, in the Wet Mountains about 60 miles (97 km) southwest of Colorado Springs, in April 1956. They occurred singly, in pairs and, on one reported occasion, in a group of three. They were said to have been about the size of basketballs, and to have pulsated. The publisher of a local newspaper was a member of a 50-strong group of people who chased a light around the cemetery one evening, to no avail: when they cornered the light it simply vanished and popped up at another spot. The behaviour of the lights varied: they would sometimes hover around tombstones, or simply sit glowing on the ground, or fly around at head height. Cemeteries are, of course, classic places for supposed 'corpse candles' – Will-o-the-Wisps – but clearly this phenomenon was no flicker of gas issuing from the ground. It stopped its appearances after a time.

PINNACLES LIGHT, CALIFORNIA

This photographed incident was uncovered by chance. I gave a talk on the subject of earth lights in San Francisco at the home of David Kubrin (who took his BSc in physics at Caltech, and his PhD in the history of science at Cornell). It was only on hearing this that he found a context for a remarkable sighting he had in 1973 at the Pinnacles National Monument, between Hollister and King City in California. This led him to seek out a photograph he had kept but never had published (Plate 20).

Kubrin and his former wife, Karen, visited the monument in late April or early May in 1973. They returned to their car in the parking lot as it was getting dusk. They suddenly saw a light streak by just above the treetops. 'It had a remarkable speed,' Kubrin told me. 'It followed no earthly laws of physics I was aware of.' Karen states: 'As I vaguely recall, we saw the path of a light object streaking and swirling across the sky. There were colours (or maybe it was the sunset) . . . I recall pinks, blues.' David Kubrin observed that the light produced shockwaves in the air that spread out from it while it was in motion. Then, *it stopped without deceleration*. It was this contradiction between an appearance of mass (the shockwaves), and the arrested motion as if the thing was weightless, that scrambled Kubrin's understanding of the physics governing the thing. The light was basically ovoid in form and, Kubrin says, 'when it stopped, or perhaps a bit before, the thing went into a spin, and the spin seemed to have the effect of dissipating its energy or light. . . . We watched as it lost its definitive shape and merged somewhat with the surrounding air.' It was as this spinning manoeuvre commenced that Kubrin took his photograph – not expecting anything to come out on the picture. Two other people in the car park came over to exclaim about the light.[9,10,11]

Kubrin's picture shows a golden light core partially surrounded by a flare of light, against a darkening blue sky. The core seems to have a lattice of light within it, but this could be just an effect of motion. It is difficult not to notice, however, that the lightform does bear similarity to the classic 'flying saucer' shape.

The Pinnacles are spectacular spire-like columns and jagged peaks produced by the action of weathering and erosion on the different kinds of rocks that occur together at this site. Beneath them is an intricate system of caves. The site is bounded on the east by the great San Andreas Fault, 30 miles (48 km) along which to the north is Hollister, where EQLs have been seen. In 1961, for example, Reese Dooley, a poultry farmer, saw a number of small, sequential flashes at different, random places on a hillside near Hollister at the time of two earth tremors.[12] The San Andreas Fault is on the margin of the North American and Pacific tectonic plates.

NORTH CAROLINA

Like Texas, this state harbours more than one earth light spot, the most famous being Brown Mountain. This is situated in the Blue Ridge country of the Appalachian Range near the towns of Morganton and Lenoir in the western part of the state. The mountain is composed of Cranberry Granite and rises to 2600 feet (795 metres). It is not known when lights were first seen on and around the mountain, though supposedly the legend of a Cherokee-Catawba battle around AD 1200 attributes the lights to the torches of tribeswomen searching for slain menfolk.[13] It was recorded in 1922 that a resident of Morganton claimed that people came to the area to view the lights '60 years ago'. A fishing party reported light phenomena in 1908 or 1909, and in 1910 a clergyman, C. E. Gregory, is generally thought to have been the first person really to bring the lights to general public notice. The first documented reference seems to have been an item in the *Charlotte Daily Observer* of 13 September 1913. This credited the lights' discovery to the fishing party and claimed that 'the mysterious light is seen just above the horizon almost every night. . . . With punctual regularity it rises in the southeasterly direction just over the lower slope of Brown Mountain. . . . It looks much like a toy fire balloon, a distinct ball, with no atmosphere about it . . . and very red. . . . The light is visible in all seasons.' The article pointed out that the light was also seen in various positions in the country around Brown Mountain. Gregory noted on one occasion that the light appeared like a ball of incandescent gas, in which seething motion could be observed. To anyone who has made a detailed study of earth lights, this little-known but characteristic quality of the lights when seen closely will be familiar.[14]

In response to the rising public interest in the lights, the US Geological Survey sent investigator D. B. Sterrett to the area. He concluded that they were misinterpreted locomotive headlights. However, during the summer of 1916 a disastrous flood swept down the Catawba Valley beneath Brown Mountain and its surrounding peaks, and trains were not able to run for several weeks anywhere in the region. Yet the lights were still seen: clearly

Sterrett's explanation could not be the sole cause of the phenomena.

In April 1916 an expedition in search of the lights was mounted by H. C. Martin and L. H. Coffey. They split into two groups. Dr Coffey's party saw three occurrences of a light over Adams Mountain, and Martin's group saw two lights flash out among the trees on the east side of Brown Mountain, a little below its summit. These lights moved horizontally southeastward, floating in and out of the ravines.

Professor W. G. Perry, writing a 1919 account of the lights for the Smithsonian Institution, said he had personally witnessed one of the lights: 'Suddenly there blazed in the sky, apparently above the mountain, near one end of it, a steadily glowing ball of light. It . . . blazed with a slightly yellow light, lasted about half a minute, and then abruptly disappeared.'

In 1921 Dr W. J. Humphreys of the US Weather Bureau decided that the lights were due to a form of electrical discharge akin to the 'Andes Light' (see Chapters 1 and 6). In a letter to a senator during the same year, the editor of the *Charlotte Daily Observer* described the Brown Mountain light as 'a pale white light, as one seen through a ground glass globe, and there is a faint, irregularly shaped halo around it.'

Growing interest necessitated further action by the Geological Survey, so George Rogers Mansfield came to the area for a fortnight to sort things out. He conducted interviews, carried out a survey of the mountains from various viewing positions and watched for lights himself, in the company of various locals, on seven evenings – 'four of them until after midnight' – during March and April 1922. He saw a range of lights, though none that matched the descriptions provided by Coffey, Martin or Gregory. Local residents who accompanied Mansfield on his skywatches were divided as to whether he had seen the real Brown Mountain light or not. But from his observations he concluded that 'about 44 per cent of the lights he saw were due to automobile headlights, 33 per cent to locomotive headlights, 10 per cent to stationary lights, and 10 per cent to brush fires.'[15] As Charles Fort later caustically commented in *Lo!*: 'Tot that up, and see that efficiency can't go further.'

Mansfield attempted to explain the Martin sighting as perhaps due to fireflies 'appearing unduly large because his eyes were focused on the distant hillside', but admitted that an entolomologist in the Division of Insects of the National Museum stated that such an explanation was 'improbable' on various grounds. Mansfield further suggested that the seething light the Reverend Gregory had seen might have been due to an illicit still on the mountainside. While these suggestions are weak, stretched points to dismiss any possibility of there being an unexplained core phenomenon occurring at Brown Mountain, there is no doubt that Mansfield did identify some lights as locomotives, and probably *some* others as car headlights. Quite a few of his identifications were assumptions on his part, however, and it seems unlikely that he witnessed the full range of reported light phenomena. Patricia Cantor has summed up the scope of Brown Mountain lights thus:

> One researcher narrowed the descriptions to four types, including 'Toy Balloon', 'Misty Sphere', 'Floodlight' and 'Skyrocket'. . . . They are very red, or slightly yellow, or white. They can appear singly, in pairs or four or five at a time. They appear near the top and then rise above Brown Mountain, or flash out among the trees. . . . They can be stationary or move rapidly, up, down and horizontally. Observers report personal reactions of feeling mildly curious to losing sleep after a sighting to being dazed. One man had such an intense reaction to the lights that he published a booklet about his experiences. In it he describes the lights as manifestations of intelligent beings who lead him deep into Brown Mountain to warn him that man is in great danger of destroying himself.
>
> During most sightings the lights are seen far off in the distance – three or four miles away. However, in one rare close encounter the lights seem as '15-inch balls of yellow or blue-white fire', that emit a sizzling noise. In another encounter, a group of researchers was 'buzzed by a sizzling ball of fire' while standing atop a tower they had built from which to study the phenomenon. After the incident the group disbanded.[16]

Mansfield's report was rejected by most local people, some of whom pointed out that automobiles could not get around after the 1916 floods any more than could trains. In 1925 it was revealed that the lights do not appear after periods of long drought, and one witness, with his back to the railroad and highway, saw 'a strange light suddenly come up from the mountain-top, and keep moving until it was lost in the sky'.[17]

In 1977 a research group called Orion suggested that the lights seen above the ridge of Brown Mountain were actually the refraction of distant city lights and not the true phenomenon. They placed a 500,000 candle-power arc light in a town 22 miles (35 km) east of Brown Mountain and aimed its beam to the west of the mountain where observers were stationed. Its characteristic blue-white beam was seen by the watchers as 'an orange-red orb apparently hovering several degrees above Brown Mountain's crest'. While it might be fair to wonder how often the good townsfolk direct a half-million candlepower searchlight into the sky, this experiment probably does help separate the wheat from the chaff in the Brown Mountain problem. Michael A. Frizzell, whose Enigma group started working with Orion on the Brown Mountain mystery in 1978, estimated that perhaps only 5 per cent of reports relate to a truly anomalous light, which, he felt, consisted of a rare type of event among the trees on the mountain.

Another earth light location in North Carolina that has obtained some fame is Maco Station, about 14 miles (22.5 km) west of the coastal town of Wilmington in the extreme southeast of the state. The light has the same legend as the Bragg Road light in Texas – the ghost of a decapitated brakeman looking for his head. The light has been recorded as appearing above a particular stretch of railway track near Maco Station since the 1860s. It always manifests about 3 feet (1 metre) above one of the rails and moves east. It starts with a flicker, then grows much brighter and begins to move along above the rail reaching considerable velocity, swinging from side to side as it goes. On nights when it is appearing, it shows itself at intervals of 15 minutes or so. At the end of its run of about 100 yards (92 metres) the light usually halts suddenly, hovers and glows, then races back along the track to its point of origin, hovers again, then extinguishes itself. It looks like 'a moon in miniature', claims writer John Harden. One witness told Harden that, 'Sometimes the ghost light is so bright that you can almost read by it. It rises up from the side of the tracks, comes towards you and disappears. You can see the reflection along the rails.'[18] There have been reports that the appearance of the lights has stopped trains at times, the drivers thinking some kind of signal is being made. In 1873, *two* lights appeared at the location for a while. They would approach one another from opposite directions.

The car headlight explanation has again been proffered for this light. This time, however, there is no doubt that it is not the answer. The light was appearing before cars were in use, before there were any paved highways in the region, and roads have been re-routed several times without affecting the appearance of the light. Moreover, an experiment was carried out to test the theory: roads were closed around Maco for over an hour one midnight, and no cars were allowed to enter the area. But the light appeared and performed its gyrations in the usual manner.

The Maco Station light appears in all seasons and weather conditions. It may not appear for a month or so, then manifest on several nights in succession. It is a marshy area, but this seems to be one occasion where no one has persisted in trying to maintain the charade that the light is a Will-o'-the-Wisp. Which is just as well, because in Part 3 we will be able to give pointers as to its primary association.

A third spot in North Carolina were aerial and possibly associated light phenomena occurred on two separate occasions is Chimney Rock Pass, where US 74 cuts through a spur of the Blue Ridge, about 25 miles (40 km) southeast of Ashville in the western reaches of the state. In August 1806 witnesses testified before a judge that they saw remarkable phenomena on Chimney Mountain. At 6 p.m. on 31 July they said that they saw 'one thousand or ten thousand things flying in the air' around the mountain, less than a mile distant. On the slopes were seen 'glittering white appearances' which the countryfolk perceived as a throng of people, moving around a rock. They eventually disappeared, leaving a 'solemn and pleasing impression on the mind, accompanied with a dimunition of bodily strength' in the witnesses.[19]

Then in 1811 two persons living near Chimney Rock Fall filed affadavits claiming that they saw a 'ghost army' fighting in Chimney Rock Pass – an alleged phenomenon which brings to mind the ghost army of Burton Dassett in England.

VIRGINIA

In this neighbouring state to North Carolina, brilliant lightballs have been observed undulating over certain rural roads around Suffolk, particularly one called Jackson Road. Here a light appears, looking like a single car headlight, sometimes 5 feet (1.5 metres) above the middle of the road, at others veering off alongside the road. In March 1951 over 300 people turned up at the spot to observe the phenomenon. The light has been seen by many witnesses, including State Troopers. In 1964 Frank Edwards reported that old folk in the region said that the lights had been known to their grandparents.[20]

LOUISIANA

A light was first noticed at Gonzales, just south of Baton Rouge, around April 1951. It appeared on a gravel road between Gonzales and nearby Galvez. Scores of people, including the local Sheriff, observed the light flitting around the road and over nearby treetops. As far as I am aware, the phenomenon still occurs. As usual, when it is approached too closely the light always blinks out – only to blaze into sight again moments later some distance away.

OKLAHOMA

Light phenomena were reported to have erupted on a ranch near Ada, 70 miles (113 km) southeast of Oklahoma City, in the early 1960s.[21] Crowds of people began to come out to the ranch to see the display of lights which were occurring, at that time at least, on virtually a nightly basis. The commotion eventually persuaded John Bennett, a reporter for the Ada *Evening News*, to visit the place himself. He saw an orange light appear through a screen of trees. At first he dismissed it as a house light, though many in the accompanying crowd knew better. The light grew larger, changing from red to yellow. For five minutes it glowed like 'a dying ember', then, as Bennett recounts, the light began to move, flickering up and down, then darting sideways:

> The single ball of light appeared about three feet in diameter. . . . During its fantastic flight back and forth it changed colors. . . . But it stayed in one general area, behind what looked like a sparse growth of trees. . . .
>
> In the tree line the light had changed colors again and was beginning to get more active. Suddenly a piece of the glow broke away and started a rapid bouncy course across the field in front of us. It looked like a luminous basket-ball, and about the same size. It danced before our eyes about 100 yards out front. We traced the glowing trajectory that appeared like a giant lightning bug, until it went out.

Bennett left after an hour, but not before he interviewed a young man in the crowd who said that a week before a piece of the main lightball had bounded across the field and had come right up to the fence where they were standing. 'I didn't move and it was like it was looking right at me,' the witness told Bennett.

THE OZARK LIGHTS

The Ozarks are a modest mountain chain, really a dissected plateau, located mainly in Missouri but extending south into Arkansas where it reaches its greatest height. A number of lights locations are claimed in and around the periphery of the Ozarks.

The best-known is probably the one which occurs in the extreme south-west of Missouri, on the borders of Kansas and Oklahoma, just south of Joplin. The light haunts a very localised area just about on the state line a few miles southwest of the small community of Hornet. When it was first reported, in the 1880s, it hovered around a particular road, but subsequently began appearing around another one about a mile away. Over the years, many people have tried to investigate the phenomenon, but with little success. The first recorded attempt was by students from the University of Michigan in 1942. One of their 'scientific' approaches, apparently, was to shoot at the thing with a rifle! All that happened was that the light went out for a moment, then reappeared. In 1946, the Corps of Engineers from a nearby army camp attempted a study of the light. Captain R. L. Loftin was in charge, and he decided it was all due to car headlights being refracted over a nearby ridge by varied air density, after carrying out experiments with spotlights. (A view shared by an investigator in 1945 also.) Apart from this not explaining the nineteenth-century reports of the light, Dale Kaczmarek[22] of the Ghost Research Society thinks they studied the original location by mistake, where the light no longer appeared. Loftin apparently later stated he thought they may have had the wrong location. According to Frank Edwards, the headlight theory was firmly hit on the head in 1962 when

> determined persons . . . *surrounded* the light as it bobbed along the road. Had it been nothing more than a reflection of distant headlights it would have been visible only to those moving directly towards the source. As it was, the elusive light could be seen from all sides when the investigators had a view of the road. In this particular case, it remained over the road until some of the party got to within twenty-five or thirty feet. Then it blinked out, and blinked back on again tantalizingly a few seconds later over a little nearby field.[23]

The light is said to appear nightly (another argument against headlight refractions which require special conditions to produce them), and in 1982 Kaczmarek, who has 'debunked' at least one spooklight, went to the area to see the phenomenon for himself. He was successful and obtained photographs (Plate 21) as well as first-class observations. Kaczmarek and his companions 'literally chased this light up and down the road from dusk to dawn, but could never catch it'. They saw the light best after 3 a.m., when the traffic had died down and they were the only ones on the road. Kaczmarek saw it through binoculars a few feet above the ground near a barn. The group of witnesses were less than 100 yards (92 metres) away and got a clear view:

> The light appeared to be a diamond-shaped object with a golden hue and a hollow center. You could actually see trees and bushes right through it. It stayed in that area for about sixty seconds and then dropped behind a hill. The area where the light was a second ago still glowed with some kind of luminosity or phosphorescence. The area twinkled with energy.

The light rose and fell another three times, and then the witnesses crept up the hill in their car hoping to see where it had gone. Before they could reach the top, the light appeared in the middle of the road less than 75 yards (69 metres) from them. It disappeared again, and when they reached the crest of the hill, they saw a light over treetops more than a mile distant.

From his experiences with it, Kaczmarek felt that the light displayed cunning – 'It seems to know when someone is getting too close. The light also seems to react to light, sound and movement.' He quoted the case of a group of other witnesses studying the light over the same period as Kaczmarek: they had been studying the light when it suddenly sprang into visibility only 10 yards (9 metres) from them. It was 'about the size of a basketball, orange-yellow in color, throbbing and slowly rolling along the ground.' Awestruck rather than frightened, the witnesses remained very quiet, fascinated by the sight. But at the split second a car crunched along the gravel road behind them, the light rose into the air, split into two, and shot off in opposite directions into the surrounding woods.

The objective Kaczmarek does not know what the light is, but his investigations convince him that it cannot be accounted for by refracted headlights, Will-o'-the-Wisp, natural phosphorescence or ball lightning.

Folktales abound for the cause of the light, including the now familiar motifs of a person looking for his head. The nearby Quapaw Indians hand down a tradition that the light is the merged spirits of two doomed lovers. This suggests some antiquity for the light, and caused Frank Edwards to comment: 'Whatever it is, it always was.'

Another light has been seen and photographed on a little-used railroad track near Gurdon, between Little Rock and Texarkana in Arkansas. It seems never to have been seen at close quarters, however, and could be a refraction mirage of distant car lights. Nevertheless, it does not show polarisation when viewed through filters, which would be expected from a mirage effect.

Writing in *Fate* (May 1982), Franklin L. Ward refers to what may be another Ozark light phenomenon. He recounts an experience he had in October 1938, but gives no more detailed location than 'a rural area of southern Missouri'. Ward was 16 at the time, and he and his father were taken to an old sawmill pond by a local fellow who told them of a 'spook' which appeared regularly next to the pond, moved around and came back to its point of origin to disappear. They watched the pond for only 20 minutes when 'a small blue light began to form near the ground' at one end of the pool. It started off 'about the size of a softball and . . . expanded to the size of a basketball' changing from 'a clear cornflower blue' to more of a 'robin's egg blue'. They approached to within 10 feet (3 metres) of it, when it rose to about a yard off the ground and began to move away. The local man marked the point of origin with a tree branch. They followed the light down a road where it adopted a swinging motion, and was undulating between 2 and 4 feet (60–120 cm) above the ground, lighting the surface of the road at its lowest points of travel. It covered over a quarter of a mile, circumnavigated a schoolhouse, passed through an abandoned log house – the observers could see its luminosity glinting through the cracks between the logs – and then made back towards the pond. At one point the lightball passed within a foot or so of Ward who heard 'a soft low hum like the 60-cycle hum of an electrical transformer'. It returned to its starting point where it hovered for about 30 seconds, becoming smaller and darker in colour, before 'settling into the sand' and disappearing. The whole sighting lasted 'for most of an hour'.

In the early months of 1973 Piedmont, in the southeast quarter of Missouri, experienced a 'UFO flap'. The wave was heralded by a sighting had by a local basketball team on the night of 21 February. They saw green, amber, red and white lights near Highway 60 acting as a unified configuration. The next night a series of sightings were made in the locality. They continued to be made over following weeks. Reports of strange lights in the sky coincided with curious TV interference throughout the whole region. On one occasion the police radio system stopped working. Domestic lighting dimmed and failed sporadically. Strange lights were reported sitting in fields and even passing under the surface of the nearby reservoir, Clearwater Lake (see the case of the Ogden reservoir in the previous chapter). A light was seen flying over the transmitter of the local radio station which, later that same evening, was 'knocked out'. By March 'UFO fever' had gripped the town, and roads around the place were lined with sightseers and parked cars. People phoned in sightings to the local radio station. There was an air of excitement. Hesitantly at first, Dr Harley D. Rutledge, a physics professor from Southeast Missouri State University, began to study the phenomena being reported. He made a preliminary field trip in early April with a colleague. His first sighting was of an unspectacular stationary light near Pyle's Mountain which his colleague was convinced was distant car headlights. Subsequent experiments by Rutledge tended to disprove this. But his first sighting of a clearly inexplicable light came while he was being flown in a light aircraft over Clark Mountain one night. An amber light appeared on the slope of the mountain near the top. The pilot also saw the light and made an immediate turn to approach it. The light promptly went out, but within seconds another popped into visibility a few ridges away. Through binoculars Rutledge felt he could see a 'slight to-and-fro' motion of the light. Quite suddenly, as they flew towards it, the light shot straight up at an incredible speed.

Rutledge went on to assemble funding and volunteers to carry out fieldwork in the flap zone and to pursue the phenomenon wherever it might occur in that quarter of Missouri over the next few years. The results of his efforts were written up in *Project Identification* in 1981.[24]

In the course of this work, Rutledge personally had 152 sightings of phenomena he found inexplicable. Of these, seven seemed to be what he interpreted as structured craft – 'disks' and 'a bullet-shaped object that disappeared in daylight' – and the rest (95 per cent) were 'more or less lights'.[25]

In all, over 700 photographs were taken by Rutledge and his helpers, and by local residents such as former photography teacher Maude Jefferis of Piedmont (see Plate 22.) Most of these show blobs of light or simply lines of light marking the course of the lights in their trajectories during long exposures.

UTAH

The rural communities within the remote Uintah Basin of northeast Utah – around the towns of Vernal, Roosevelt and Duchesne – experienced an intense period of UFO sightings between 1966 and 1968, with a lower incidence of activity reported both before and after those years. East to west, the Uintah Basin is approximately 75 miles (121 km) long, and extends 30 miles (48 km) north to south. It has an elevation of over 5000 feet (1529 metres), is bounded on the north by the high Uinta Mountains, and has desert country to the south.

A local high school science teacher, Joseph Junior Hicks, became a focus for people who saw strange light phenomena, and he began to keep files of the reports he received and interviews he conducted. These later became the basis for an excellent study carried out by biologist Dr Frank Salisbury, of Utah State University.[26]

The main type of phenomenon observed was a large light, its colour variously described as like 'fire', the glow of a 'setting sun', or the amber colour of a rising 'harvest moon'. Indeed, the likeness of the lightform to a ruddy full moon was frequently made by witnesses – a typically rural comparison. It was rarely fully spherical, however. Its most common form was a sort of hemisphere with a flat bottom. But other kinds of lights were also seen, and occasional odd shapes were reported, including 'rockets' and 'cigars'. Infrequent daylight sightings of spheroid or discoid silvery objects were made. No 'entities' were reported. Certain locales within the area seemed specially haunted by light phenomena – a dry desert gulley known as Halfway Hollow between Roosevelt and Vernal is one example. Some of these seem to be genuine 'ufocals', but others may have been the effect of Junior Hicks's circle of activity within the Basin. Many sightings had multiple witnesses, with it not being uncommon for three or more observers to be involved – one lightform was seen by at least 40 people! There were many instances of close views of lightforms, some of which were in sight for relatively long periods – up to an hour in a few cases. A full cross-section of the local population reported light phenomena, including community and church (Mormon) leaders, teachers, hunters, farmers, housewives. Age groups of witnesses ranged from children to the elderly. The lights were usually silent, but on close encounters 'whistling' or 'humming' was sometimes heard. They were extraordinarily manoeuverable, performing unbelievable aerobatics, and often disappearing to the far distance in a fraction of a second. Most of the lights were estimated to be tens of feet across.

Salisbury made a careful analysis of moon and planetary positions for the years under study, and finally dismissed all but 80 of the 300–400 sightings that had been filed. These 80 cases, which he felt could not be readily explained away, involved 260 witnesses out of a population of only about

4000 people (excluding the small towns of Duchesne and Vernal) in the area most affected by the light phenomena. Salisbury's investigations were to show that, on further questioning, many of these witnesses had had what they believed to be other sightings, but which they had not reported. Salisbury also gathered that the Indian communities in the region had seen many lights, but were not keen on reporting them. Certainly only a few Indian witnesses figure in the files.

The vast majority of the sightings were of lights. Some of the aerial phenomena projected curious 'beams of light'. But Salisbury noted that the reports only indicated that the ground lit up if the beam actually touched the ground, otherwise the ray of light seemed to 'stop' in mid air. In other cases, lightbeams appeared as if 'spliced' with varying intensities of light. In September 1966, schoolteachers Valda and Donna Massey saw a red light 'as big as a house' projecting one of these curious beams of light horizontally. Other colours of light could be seen in the main shape. It settled 'real soft' onto a hilltop northeast of Vernal. A Mormon bishop, Garth Batty, was called and he and the Masseys watched the light on the knoll for about an hour. They finally decided to get in a car and approach it. But as they started to do so, the light suddenly rose vertically so fast it 'just vanished'. About three nights later a huge yellow light was seen to land on the same knoll. One witness said, 'It didn't shoot out any rays from it, and it was a different type of light from what our lights are.' Later that same month, Joe Ann Harris was driving south from Roosevelt with an Indian woman and four teenage Indian girls when a 'real big' brilliant yellow light came at them, sloping down through the night air until 'it got real close'. The Indian girls crouched on the car floor 'screaming and hollering'. It transpired that the light was seen by another witness on a nearby road at the same time. In an incident a year later, September 1967, a group of hunters on the road to Buck Canyon from Ouray saw a large truck rolled onto its side in a gully by the road. Looking around, they could not find a sign of the driver. The hunters set off again on the journey when the driver, Lee Albertson, saw what he took to be a fire, out in the night to his right. He pulled up the vehicle, and saw that 'this son-of-a-gun was moving'. They all watched the light as it went far up into the sky, then descended to another location. It looked like 'a big harvest moon' and 'it kind of sat there like it was watching'. Then it took off again and 'it got right out of there'. Albertson told Salisbury during his interview about something else that he had 'often wondered about'. He recalled one evening when 'the northern lights, believe it or not, were very much in evidence. This was the first time I've seen them, but it was really something. I mean, it wasn't just vaguely; it was outstanding.' While the fiery light he and his colleagues had seen bore no relationship to these 'northern lights', Albertson could not help but wonder if they had 'something to do with the phenomenon'. The truck was recovered some time later, and no reason was given as to why it had come off the road. Salisbury noted the obvious possibility that the driver had seen a similar light to the hunters and the shock had caused him to crash.

Lights appeared near reservoirs and power lines as well as alongside roads and over the desert and mountains. Salisbury's study of the witnesses led him to note that while they may be a rural people, as farmers and deer hunters they were exceptionally well trained at estimating sizes and distances. No photographs were presented, but, as Salisbury observed, these outdoor people used cameras only for family events, and did not carry such equipment around with them. Salisbury soundly dismissed the notion that the core of sighting reports could be the result of hoax, conspiracy and lies. Apart from the calibre of witnesses, the numbers involved, and the homogeneity of the accounts, none of the observers sought publicity. The biologist had to seek them out, and some would not talk to him until he showed identification that he was a *bona fide* academic. Something extra-ordinary had certainly happened.

Salisbury further noted that the Uintah Basin reports carried a large number of examples of lights following vehicles; 25 to 30 per cent of the reports related to lights that were on the ground at some point while they were being observed, and that in about 15 per cent of the reported sightings binoculars were used. No physical traces were claimed, nor were there any claimed electromagnetic effects, such as car engines or lights failing. One curious place was identified by one witness, however, where car headlights

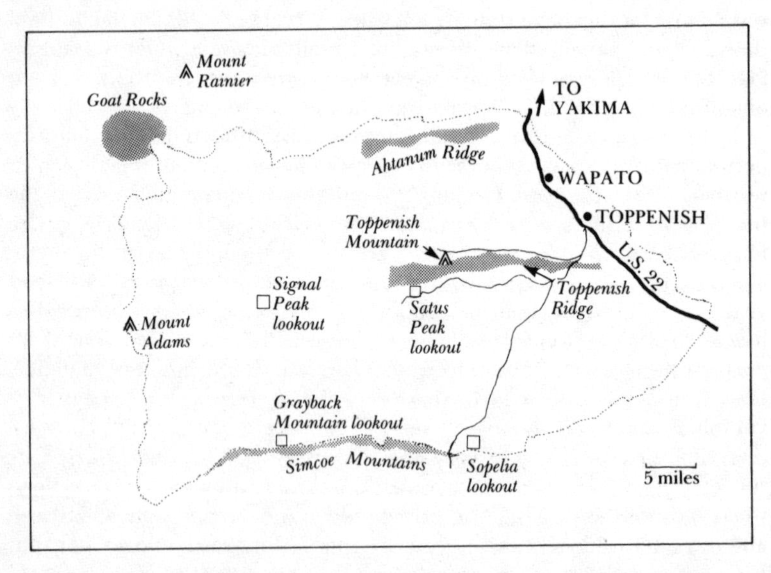

FIGURE 11 Basic layout of the Yakima Reservation area of Washington State, USA. (*After Greg Long.*)

would blink on and off for a number of yards, but this does not seem to be readily associated with a sighting.

Finally, Salisbury noted some reports of mild psychic phenomena during the most intense period of sightings, and remarked that some witnesses felt that the lights reacted to them.

YAKIMA, WASHINGTON STATE

We have been here before in this book: it was the airfield at Yakima where Kenneth Arnold first landed after seeing the nine discs near Mount Rainier in 1947 (Chapter 2). In the 1970s, the general area was again visited by strange lights. The observers this time were, in the main, fire control officers situated on the Yakima Indian Reservation, a 50-mile-wide (80 km) area located immediately east of the main Cascade range, between Mount Adams to the west and the town of Toppenish to the east (Figure 11). The reservation is about 130 miles (209 km) southeast of Seattle.

Lights have been reported occurring over the 3,500 square miles of the reservation and in areas adjacent to it. The main period of activity seems to have been between 1972 and 1978, though remarkable sightings were reported into the early 1980s, and sporadically from before 1969. The outbreak was at its most intense between 1972 and 1974 when activity on the reservation 'was heavier than that of any other area in the United States at the time – although the high level of activity could be the 'result' of a greater awareness and conscious documentation' on the part of the fire officers.[27]

A considerable portion of the reservation is timbered, and there are permanent fire observation stations positioned throughout the area. Fire wardens are trained observers, who know every nuance of the landscape they scan visually and almost constantly: it is essential that this is the case, for they must be able to spot the first hint of a fire yet without causing false alarms. All the fire lookouts who reported light phenomena had many years' experience studying the Yakima Reservation, and some have, indeed, Yakima Indian lineage.

Nocturnal lights formed the vast majority of sightings, and these usually in the form of red-orange or yellow-orange balls of light floating over various locations – particularly Goat Rocks just to the northwest of the reservation boundary; north of the Simcoe Mountains which form the southern edge of the reservation, and amongst the buttes and valleys of the western part of the area. These type of lights were seen mainly in the early autumn months, and never more than two were seen together at one time. Radio links between the fire lookouts allowed the position of some lights to be triangulated. These globes of light were usually quite large, though smaller versions, the size of 'ping-pong balls', were seen bounding along the ground on Satus Peak and along Toppenish Ridge.

But more complex lights also were reported – white lights with smaller, multi-coloured lights on them; curiously shaped lights, and so on. Columns and flares of light were reported as well. These phenomena looked like fires, and obviously caused the lookouts special consternation. They could look just like a tree on fire, though it was obvious that if such was the case, the conflagration could not be localised to a single tree. When these triangulated positions were checked by aircraft in first daylight following the sightings, no burning or scorching was ever discernable.

As usual in any wave of sightings, apparently structured 'craft' occurred only very occasionally, compared to light phenomena. Here they were the ubiquitous flying or floating disc, and the equally common tubular, 'cigar-shaped' or 'rocket-shaped' forms.

A particularly interesting aspect of the Yakima wave was the occurrence of abnormal meteorological features, phenomena sometimes noted at other earth lights hotspots. Extraordinarily slow 'meteors' were seen. Unusual clouds were spotted: in one case, for example, fire lookout Dorothea Sturm was awakened one moonless night in her Satus Peak observation station by a bright glow. Looking outside she saw a roundish cloud 12 miles (19 km) away alternately brightening and dimming with white light. It finally faded from view. On another occasion, fire control chief Bill Vogel saw a moving bright orange light that split into two and went out. Moments later, a pin-point of light was seen dropping towards the ground near Satus Peak. Vogel momentarily took his eyes off the area and when he looked again there was a white, glowing cloud where the light had been. The cloud began moving, dissipating as it went, leaving a phosphorescent effect behind it. Other 'meteorological' effects included curious 'beams' of light. These sometimes seemed to spring out of the sky and shine into canyons or onto trees; at other times weaker light beams seemed to project into the sky. There were also glows. These appeared over the slopes of mountains – Mount Adams on 20 July 1974, for example – and behind buttes. Glowing patches in the sky were sometimes observed: a guard at the Mill Creek Guard Station, for instance, saw a glow travel overhead at night. Then there were odd flashes of light. Former fire warden Gladys McDaniel observed blue flashes on one occasion from the Sopelia lookout tower. First, the sky took on a pale green hue, and then formed a 'big curtain' effect. The flashes then started on various points of this 'curtain', which sounds like a specialised aurora effect.

In addition to the aerial and luminous phenomena, observers also heard sounds – deep 'rumblings' from underground. Two Washington State geologists who visited Dorothea Sturm at Satus Peak apparently told her that they too had heard such sounds in the mountains of Montana, but offered no explanation.

Police officers patrolling along Pump House Road near Toppenish Ridge within the reservation reported lights chasing their cars, with the inevitable failure of car lights and engine taking place. There were also inexplicable shut-downs of the entire reservation two-way radio system.

And, there were reports of what might be termed 'poltergeist' phenomena. Fire lookouts claim to have heard voices calling in inexplicable circumstances, 'happy little voices singing', a woman screaming, someone 'hollering'. There were strong, repulsive odours. And the secretary of a wildlife biologist stationed at a government dwelling near the Pump House Road frequently heard the gravel outside crunching as if being walked on – but there would never be anyone around to make the sound. Bill Vogel also heard the crunching gravel sound at the location. The fleeting appearance of a seven-foot-tall, wild-looking humanoid was also reported.

Finally, it was the distinct feeling of many of the observers that the lights interacted with them in curious ways. 'I'm convinced these things know when you're talking about them,' Dorothea Sturm commented.

ALASKA

This state, hard by the Arctic circle, boasts one regular earth light effect – the Iliama lights. The mountains near Lake Iliama, in southwest Alaska, can produce a brilliant glow sometimes visible from 40 miles (64 km) away.

NEW JERSEY

Way across North America from Alaska, on the east coast, the 'Garden State' has an earth light near Washington Township, 40 miles (64 km) west of Newark. The 'Hooker light' is a sphere about 18 inches (45 cm) in diameter, and used to be seen swaying along an old railroad track. The legend associated with it should by now be a familar one: the light is the ghostly lantern of a dead brakeman looking for his arm which was severed along that stretch of track.

In 1976 a group of researchers calling themselves Vestigia started studying the phenomenon. Over a period of time they observed, took infra-red pictures and environmental measurements. Close-up views of the light revealed a 'bullet-shaped core' within the lightball. Electrical resistivity readings changed in the rails when it bobbed along above them. On one occasion when it 'went out' a surge of readings was obtained on a geiger counter shortly afterwards for a brief period, probably indicating radon gas emission. Low-frequency radio signals were detected radiating from the light.[28,29] The light was successfully captured on infra-red film – a shot taken on 22 April 1977 clearly shows a light down the track from the camera position. On 27 September 1977, however, an infra-red test shot was taken of two Vestigia members walking down the track. No one saw it with the naked eye, but the ghost light was recorded on the film.[30]

Specific appearances of the light have been followed by small earth tremors a few days later, and the long-term incidence of the light seems to reflect activity in the Ramapo Fault system.

When the rail track was pulled up, the light continued to appear. It had been thought that the rail was acting as an electrical conductor for the light, but it still moved along its usual course, with just a modification made to its undulations.

Some people have experienced an irrational feeling of 'dread' in the area where the light manifests.

NEW YORK STATE

One small, highly localised earth lights area in this state was brought to my attention by author Phyllis M. H. Atwater. On 5 October 1986 she was persuaded to be taken to Pine Bush, about 60 miles (97 km) northwest of New York City, to look for UFOs by a photo-journalist who had been 'tracking' aerial phenomena for several years and who claimed that the Pine Bush area was prone to such sightings. Atwater was accompanied by two women companions. The guide was convinced the lights were alien craft, and that they were seeking and mining rare minerals, with which the area around Pine Bush was supposedly rich. In the evening, she led Atwater and companions to a small side road between farms and fields, where they were treated to a remarkable light show:

> Globes would 'bubble' up . . . from behind trees, then glide sideways at tree-top height . . . then either continue rising . . . or change direction and glide back. . . . All globes or light balls were large and seemed in no hurry. They would first appear as pure white, until such time as they 'cut loose' from the tree-tops, then they would turn blood-red. Colors were pure. The lights were steady. After attaining some height in the sky, and moving closer to us, the globes would again change color, this time to green, which allowed them to almost disappear in the black sky. I heard a high-pitched whirring sound as they got closer to us, but no regular aircraft sound was made. When the light globes turned green, flashing lights were visible all around them, and the

> globe then appeared more triangular in shape. They would proceed to the bank of trees behind us, then disappear below the ridge. There were dozens of them.[31]

The women tried to get closer to the apparent source of the lights but were warned off by a local farmer who said there were coon hunters with guns and dogs in the area and it would be dangerous to proceed. He acknowledged the lights, saying they were a regular occurrence, and paid them no attention. They drove around to find another location, but the 'show' was coming to an end. Atwater felt she could not 'in all good conscience' say the lights were UFOs, but nor were they any kind of craft she had seen before (and with her former husband being a pilot she is fairly familiar with aircraft). She feels the matter deserves some serious attention.

What may be another location is the hilly, wooded, suburban area along the Hudson Valley immediately north of New York City. An outbreak of reported phenomena took place here between 1982 and 1985, stretching from White Plains in the south to Newburgh in the north, and east into Connecticut. It seems to have peaked on 24 March 1983. Fortunately, a number of local Center for UFO Studies (CUFOS) investigators were at hand, and they processed many hundreds of reports. (Their conservative estimate for the total number of reports received by police, themselves and others is put at 5,000 for the period.) Thus from the very beginning of the flap, people were on hand who were very aware of the usual traps of misidentified mundane objects (the IFOs or Identified Flying Objects), and the problems evaluating witnesses' reports. While they dismissed many, they were still left with hundreds of reports to follow up with interviews. It was a huge undertaking. They had so many reports to choose from that they could indulge in the luxury of selecting the most credible witnesses to interview – mainly people with professional and technical backgrounds, as well as police officers, security personnel and the like. Photographs and videos were taken of the phenomenon by some witnesses. J. Allen Hynek visited the scene on several occasions and took part in the investigations. He and two of the local investigators, Philip J. Imbrogno and Bob Pratt, wrote a book, *Night Siege*, on the situation.[32] Hynek died during the course of the work.

What was seen, essentially, was a configuration of lights forming a 'V' or 'boomerang' shape. The lights were multi-coloured, though sometimes just one or two colours predominated, or brilliant white lights were seen. The configuration was so tight that the lights appeared to be on an object of some kind. Many people did claim to see a structure linking the lights. Invariably, this was seen as a very dark, totally non-reflective material. Many witnesses of this structure said it was comprised of struts and joints, like bridge lattice-work. Other witnesses, though, claimed they were unable to see anything solid between or behind the lights. On some nights of high incidence, more witnesses would report a structure than those who did not, but on other nights the proportion would be reversed. One person said there seemed to be a tenuous structure, but it appeared 'ghostly'. The configuration or object would turn without banking, as if it were 'on a wheel'. Estimates of size ranged from 27 feet (8 metres) across to over 300 feet (92 metres). (It appears, though, that there were two main kinds of object, with one being smaller than the other.) It hovered or moved slowly on most occasions. In two reported instances, it floated at the pace of a walker and a jogger respectively, and one person who moved underneath it in a car said it was travelling no faster than 10 mph (16 kph). It often descended to below 1000 feet (306 metres) – frequently only a few hundred feet. Yet despite its massive size and low level, most witnesses said it was silent or gave off only a quiet humming sound. The configuration or object paid great attention to reservoirs, lakes and other bodies of water. Over Croton Falls Reservoir, for example, it was seen to send down an intense beam of white light. The thing would hover over or pass slowly across busy roadways, and on more than one occasion brought traffic on the Taconic Parkway to a halt as people gawped at the airborne wonder. Many people likened it to the giant UFO shown in Spielberg's *Close Encounters* movie. The configuration or object would sometimes send down its beam of brilliant white light onto a car, to the great consternation of the occupants. On one of these occasions a child in the car complained of a 'tingling' sensation.

The huge size of the configuration, and its apparent 'nuts and bolts' construction as perceived by some, is belied by its silent or quiet nature and its slow, floating, movement. Furthermore, the lights sometimes changed their position instantaneously, and on several reported occasions one of the lights would move off on its own some way from the main set of lights. On a few occasions the whole thing simply vanished suddenly, or disappeared in one place to reappear instantaneously nearby.

On the other hand, the length of time the object was in view (often over five minutes), the repeat performances it gave over a long period of time, the fact that it *usually* moved off to disappear from sight like a normal vehicle, are not characteristic of earth lights, nor even of apparently structured UFOs. On this basis, it could be argued that an actual craft, perhaps a secret military 'stealth' machine, was being tried out, to see how well it could handle itself over populated areas with intense air traffic control. But this seems highly improbable in itself, and a great many witness accounts would have to be dismissed.

But the sheer number, range and quality of witnesses leaves no doubt at all that *something* odd was seen in the Hudson Valley skies over the period. No one questions this. Various authorities made the suggestion that a group of rogue pilots were flying light aircraft in formation as a hoax. At one point, this definitely did happen, but some of those who had witnessed the actual phenomenon also saw this hoax formation and claimed the two effects in the sky were quite dissimilar, apart from the fact that the planes made considerable noise. Another 'official' suggestion was that microlite aircraft were doing the same thing. But the configuration of lights was seen on very windy nights, when it would have been impossible for the steady relationship between the lights to be maintained by light craft in formation, and when such a hoax would have been tantamount to a suicide attempt. Moreover, the lights of the actual phenomenon were of such intensity – lighting up the ground beneath in some cases – that they were clearly other than aircraft lights. In any case, however skilful the formation flying, it could not reproduce the slowness and manoeuvres of the V-shaped object.

The CUFOS investigators checked every possible rational explanation they could think of, and found all of them wanting.

Reports from witnesses through the entire period claimed that the lights *interacted* with them.

6 LIGHTS WORLDWIDE

Every country in the world probably has a number of locations favoured by terrain-related lights. Knowledge of them, of course, depends on whether they have been observed by anyone, how widely that information has been broadcast, in what language, and so on. It is obviously impossible, therefore, to provide a comprehensive global survey of earth lights zones: all that can be done in this chapter is merely to provide examples to illustrate the widespread and generally homogenous nature of the phenomenon we are studying.

NORWAY

Approximately 74 miles (120 km) southeast of Trondheim is the valley of Hessdalen, fairly close to the Swedish border. It is just over 7 miles (12 km) in length and the town of Røros is nearly 19 miles (30 km) southeast of it. The valley has a population of around 150 people, most of them living in isolated farms along unmade tracks. For a few years after November 1981 this remote location put itself 'on the map' by providing one of the most remarkable outbreaks of light phenomena reported anywhere in the world.

Lights appeared within or immediately local to the valley: they would spring into existence near rooftops, or hover just above the ground. Mostly, however, they were seen just below the summits of nearby mountains. Their shapes included spheres, 'bullet-shapes' with the pointed end downwards, and inverted 'Christmas tree' forms. Colours were predominantly white or yellow-white, though other colours sometimes appeared – particularly small, flashing red lights on the top or bottom of larger white forms. The lights could essentially be divided into three types:

(i) Strong, localised white or blue flashes in the sky.

(ii) Yellow or yellow-white lights. Seen within the valley, both low to the ground and high in the sky. They displayed a wide range of behaviour, being able to remain stationary for up to an hour, move slowly, or show sudden acceleration. They usually moved on a north–south course.

(iii) Configurations of lights, usually yellow and white, sometimes with a single red light. These also tended to move on a north–south course, and did so slowly. They would be seen moving near the tops of the mountains. Because the lights kept fixed positions to one another, it appeared as if they were on a dark, invisible object. Hence locals tended to use the term 'the object' when referring to this kind of sighting.[1]

Lights were observed at various times in the day, but usually in the evenings or at night. Most reports occurred in the winter: this could be because at that latitude there is very little darkness in summertime.

In March 1982 Norway's leading UFO group, *UFO-Norge*, held a meeting at nearby Ålen to which 130 locals came. From a survey taken at the meeting, it was revealed that 30 people there claimed sightings of all kinds of lights, but mainly of yellow globes.

At the end of March the Norwegian defence department sent two air force officers to study the situation. They did not see any lights, but took the reports seriously, feeling the accounts they heard to be credible. They also claimed that 'The people of Hessdalen have been seeing luminous objects since 1944, but many years have passed before they dared to talk about the sightings.'[2]

But this seems to have been the only official interest in the lights in the early part of the outbreak. By the summer of 1983 hundreds of reports of light phenomena had been made. People also complained of curious sounds at Hessdalen: in 1981 a sound like 'a train passing beneath them in a tunnel' were reported. 'Banging' sounds like thunder were also heard, and in 1984 'deep booms' were audible. Many of the sounds seemed as if they were moving around the mountains.[3] Unusual interference with TV and radio reception was also experienced. Some kind of organised approach to studying the situation needed to be made, and in lieu of official interest, Project Hessdalen was formed in June 1983 by co-operation amongst Norwegian, Swedish and Finnish UFO groups. The five directors were Leif Havik, Odd-Gunnar Røed, Erling Strand, Haken Ekstrand and Jan Fjellander. They raised interest amongst ufologists internationally, provided or borrowed various items of equipment, and raised modest financial support from well-wishers. Field operations got under way on 21 January and ended on 26 February, 1984. Apart from the Missouri work carried out by Harley D. Rutledge, this has been the only time that equipped field teams have been present to study luminous phenomena as they occurred in the landscape.

They had to work in appallingly difficult winter conditions – with temperatures as low as −30°C. Their main field base was a caravan equipped with electricity, where most of the equipment was housed (Plates 25 and 26), but there were also scattered field observation positions around the area. The Project finally had use of a magnetometer to measure any changes in magnetic field; an infra-red viewer; a spectrum-analyser to study possible radio emissions; a seismograph to check movements in the Earth's crust at Hessdalen; a radar unit; a laser; geiger counters, and a variety of photographic equipment including cameras with lens gratings to show up spectra of any lights that might be photographed. Volunteers were mustered to give as much cover as possible throughout the field period – the maximum number involved overall in the Project was 40. It was a noble and brave undertaking.

The outcome of this period of the work was enough to confirm that a genuine anomalous phenomenon was taking place in and around the valley, without providing anything definitive, which was not really expected. During the field period, 188 lights were reported. Some of these were definitely established as aircraft lights, while others were equally certainly some unknown phenomenon. Photographs were successfully taken of unknown light phenomena (Plates 27–29), but out of seven films taken through the special lens gratings, only four pictures came out well enough to show spectra of lights, and only two of these were strong. One spectrum of one 'high strangeness' object was analysed and showed a wavelength range from 560 nm (nanometres) to the maximum the film could respond to – 630 nm. The radar provided 36 recordings that were not considered radar 'angels' or artefacts. Of these, three were also 'probably seen as lights' at the same time; such cases are known as 'radar-visuals' in ufological jargon and are highly prized. A baffling observation was made when, in one of these instances, a light that was constantly visible to the eye only returned a radar echo every second sweep of the radar dish. In another visible case, the radar measured the speed of a silent light at 8500 metres per second. The spectrum analyser did not register anything unusual while lights were being seen, but odd readings were obtained at times during the field period. These showed up as 'spikes' at approximately 80 mHz (megaHertz). Neither the geiger counters nor the infra-red viewer rendered any information, but the Hessdalen team felt this could simply be because the equipment was too far from any of the light phenomena. Magnetic field changes were recorded by the magnetometer in apparent relationship with about 40 per cent of perceived lights, but one high strangeness light did not cause any unusual magnetic registration. It is possible that the variations in magnetic field were simply coincidental. The laser was used to point at a flashing light on two occasions. In the first instance, on 12 February, 1984, a light moving slowly north was seen. It was flashing regularly. The Hessdalen researchers pointed their laser at it, whereupon it immediately changed its flashing sequence. The laser was stopped, and the unknown light's flashes returned to their former periodicity. The laser experiment was repeated four times and in each case the light apparently responded by changing its flashing sequence. The second case occurred later the same day, when a similar apparent response pattern was obtained four times out of five. The test was carried out with the observer of the light being unaware of when the laser was being directed at the moving light.

The seismograph was installed in late October 1983, but the Hessdalen team's close observations of its recordings seem to have been made only between 1 and 24 February on the basis of data released. Due, it appears, to a misunderstanding in interpretation, the Hessdalen team in their 1984 technical report failed to see any connection between the seismograph readings and the incidence of lights for the period. As a member of the Project's advisory committee, I was able to show that there was, in fact, a strong correlation, as we shall see in Part 3 along with other geological observations regarding Hessdalen.

The Hessdalen team learned much in the 1984 field period, particularly regarding changes and improvements in equipment and techniques. Members of the team also saw light phenomena themselves. Perhaps the most remarkable was the observation by Leif Havik of a small red light flashing around his feet on 20 February as he left the caravan HQ. There was another witness. They felt that this strong light could be a patch of luminosity projected from above onto the snow. They provide no evidence for this, and this interpretation seems to be largely based on the similarity between the light's colour and that of their laser beam. Havik also saw an oblong configuration of lights, which he was able to photograph, passing slowly and silently in front of a mountain.[4]

With the help of the Norwegian Meteorological Institute, the Norwegian Institute of Scientific Research and Enlightenment (NIVFO) set up a meteorological station at Hessdalen during April 1984. This was to test earlier suggestions that the lights might be due to temperature variations in air levels causing the refraction of distant light sources. NIVFO's findings suggested that some lights could be explained in this way, but that a residue was left that was mainly comprised of large lights that moved in undulations and which in daylight often appeared like metallic balls or discs.

J. Allen Hynek visited Hessdalen and met with the Project team leaders in 1985. He felt that 'we have something important in Hessdalen'.

Project Hessdalen carried out further work in the area during the winters of 1985 and 1986, but although lights were reported from time to time, it is clear that the peak of activity occurred between 1982 and 1984. Nevertheless, the phenomena are still present to some extent. In 1987 Leif Havik made a private trip to Hessdalen and in January he saw 'an oblong yellow-white light' in the mountains.[5] Other people reported lights during his stay in the area.

In November 1981, immediately before the onset of the activity at Hessdalen, another Norwegian location provided an outbreak of light phenomena. This was at Arendal, a popular coastal town on the southeast coast. UFO-Norge obtained 78 successful photographs from this wave, in which many people reported phenomena. Although structured craft were reported by witnesses in some cases, the photos show only lights, though lights of complex configurations and displaying remarkable motion.[6]

SWEDEN

Swedish researcher Dan Mattsson has made a detailed study of reported UFO activity over Sweden. His findings lead him to identify seven 'window' areas in the country (Figure 12).[7] Some exhibit above-average numbers of sightings on a fairly regular basis, while others have had outbreaks of phenomena for limited periods. The areas as numbered in the figure are:

1 *The Gislaved-Gnosjo district* in southwest Sweden. Here Mattsson found 3.6 times the average number of Swedish sightings until an earthquake there in September 1976. A sighting was made on the evening of the quake, but very few have been reported subsequently.

2 *The Kolmarden area* between Norrkoping and Nyköping, on the east, Baltic, coast south of Stockholm. This yields four times the expected number of observations, and has had a constantly high rate for 20 years. The following examples are merely a brief, illustrative selection of the types of sightings that have haunted the area.

In April 1972 motorist Sven Närholm was followed for over ten minutes by a blue-green ball of light near Nävekvarn. The metre-wide lightball came to within 11 yards (10 metres) of his car and apparently caused interference on the car radio. The silent sphere of light then flew to nearby woods and disappeared. In August that same year, Olof Fredriksson noticed a blue-tinged ball of light hovering over Lake Langhalsen, northwest of Nyköping, after the lights in his summer cottage dimmed. He went out to observe the object. When it disappeared, the cottage lightning regained normal functioning. On 8 January 1977 at least 50 people in the Nyköping–Oxelösund area saw five red, glowing discs or balls of light. In April that year a plate-like object was seen hovering over 300 feet (100 metres) above a water tower at Oxelösund. It was over 13 feet (4 metres) in diameter and emitted a whistling sound. The object finally accelerated to Braviken Bay. This was followed by a 'strong, very dull, rumbling sound which made window-panes rattle', heard several times over a wide area. A blue-white egg-shaped light

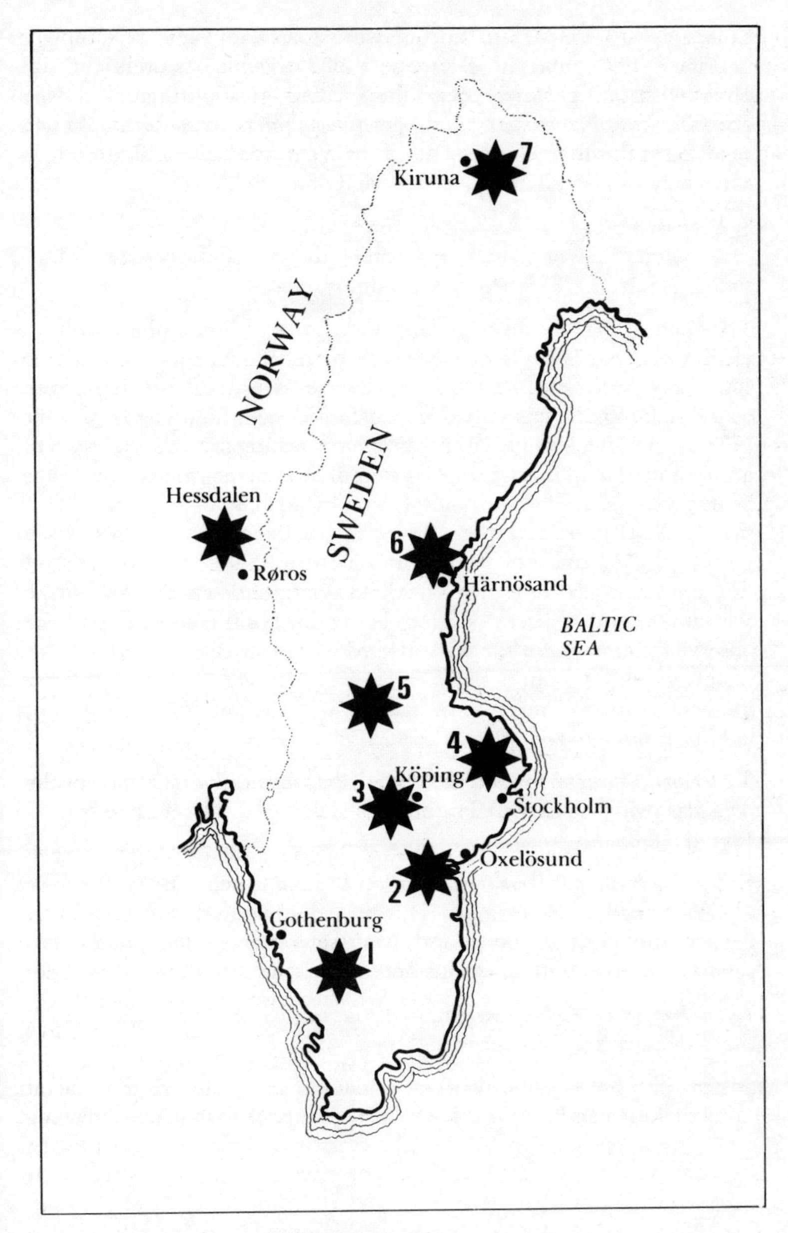

FIGURE 12 Key 'windows' of light phenomena in Sweden, according to the research of Swede Dan Mattsson. The numbers refer to the zones described in this chapter. The location of Hessdalen in neighbouring Norway is also shown.

caused a car's engines to stop between Oxelösund and Nyköping in February 1979, and in December 1982 a white 'searchlight' was observed floating slowly near the ground at Hultstugan, between Kvarsebo and Navekvarn.[8] Odd peripheral phenomena have also been reported in the area, like large numbers of car windscreens shattering inexplicably in a small area over a period of a few hours.

3 *Köping, Arboga and Kungsör district*, 80 miles (129 km) west of Stockholm. This area had eight times the national average of UFO sightings until 1978, when reports almost ceased.

4 *Vallentuna*, in the county of Uppland, north of Stockholm and in the vicinity of the ancient, sacred town of Uppsala. This area yielded around 100 reports between 1973 and 1974 – Sweden's most intense 'flap'. Some of the sightings in this period are of considerable interest. In October 1973 three 'pillars of light' were observed reaching from the clouds to the ground, then searching from the ground towards the clouds. These light beams were parallel to one another and inclined at an angle of approximately 30 degrees. As they disappeared, there were flashes on the ground. In February of 1974, four yellowish light beams were seen to 'meet horizontally' in an open field. They descended and 'went out'. In the same month a white, shining light with multi-coloured edges was observed moving across a field. Beams of light were seen coming from the object to the ground, *and from the ground towards the object* at different moments. Another instance of light beams reaching up to an airborne lightform occurred in March 1974.

5 *Dalarna*, a county roughly 120 miles (193 km) northwest of Stockholm. This has twice the expected number of sightings – a recent flap being in January 1985.

6 *The Sundsvall and Härnösand area*, on the Baltic coast about 300 miles (483 km) north of Stockholm. This provides two and half time the expected number of sightings, and, Mattsson observes, the quality of the reports are high, with fewer instances of misidentification.

7 *The Kiruna and Gallivare region*, in the far north of Sweden. This has four times the average number of sightings.

Mattsson has noted a number of geological and seismological factors relating to these windows, which we shall come back to in the next chapter.

WEST AFRICA

The earth lights riches of the vast African continent must be exceptional, but virtually no work has been done on them – Africa has more pressing problems. The following case is offered simply to represent this remarkable part of the Earth's surface, a case that hints at a location similar to European and North American spooklight haunts, and also at the absorption of such phenomena into the folklore, the psyche, of the local people.

In 1895 Mary Kingsley was on an exploration around Lake Ncovi between the Ogowe and Rembwe rivers, Gabon. She went alone one night to bathe in the waters of the lake when she noticed an unusual phenomenon:

> Down through the forest on the lake bank opposite came a violet ball the size of a small orange. When it reached the sand beach it hovered along it to and fro close to the ground. In a few minutes another ball of similarly coloured light came towards it from behind one of the islets, and the two wavered to and fro on the beach, sometimes circling each other. I made off towards them in a canoe, thinking – as I still do – they were some brand new kind of luminous insect. When I got onto their beach one of them went off into the bushes and the other away over the water. I followed in the canoe . . . and, when I thought I had almost got it, it went down into the water and I could see it glowing as it sunk until it vanished in the depths.[9]

Mary Kingsley would have been disappointed to know that nearly a century later no such exotic luminous insect has ever been discovered, and surprised, perhaps, to learn that what she saw was experienced by others near bodies of water throughout the world. Her account is particularly interesting for the observation of the light still glowing beneath the water. There have been reports of such phenomena from lone transatlantic sailors.

Kingsley later asked local people what they thought of the phenomena she had witnessed, and they explained to her that such a light was considered by them to be an *Aku* or devil.

INDIA

The comments for Africa apply here to the Indian sub-continent. Again, without a separate research programme, a comprehensive set of earth lights locations for the country cannot be given, but the 'library angels' have conspired to throw up two entirely independent reports during my research that allow one Indian earth lights lair to be positively identified for the first time. The area is around Darjeeling, in the extreme northeast of India in the Himalayan foothills. The first witness is a correspondent to *Folklore*, writing around the turn of the century:

> I was staying on a tea-garden near Darjeeling last year, and one evening as we were walking round the flower-garden . . . our eyes were caught by a light like that of a lantern being carried down the path which leads to the vegetable garden 200 feet below. My host sent for the 'Mahli' . . . 'Oh,' said the man, 'that is one of the "chota-admis" ' (i.e. little men); and on being asked to explain, he said these little men lived underground, and only came out at night. He did not appear to be very clear as to what their occupation was, but they always walk or fly about with lanterns. They are about three feet high, and they will never allow anyone to get near them; but if by any chance one was to come upon them unexpectedly, they would quickly disappear, and the person who saw them would soon become ill and probably die. . . . Whilst he was speaking we watched the light, which apparently left the path, and in two or three minutes flew across to another portion of the hill . . . which would take an ordinary individual at least half an hour . . . then it disappeared, and we saw no more that night, but two or three times afterwards we saw similar lights, sometimes . . . along the paths and at others flying across the dips in the hills. . . . The light was too large and not erratic enough for any firefly . . . more like a lantern than anything else we could think of.

The correspondent wondered 'is it folklore, or is it in the province of physical research, or of natural history?' then added with a breadth of thinking that is so frequently lacking in today's scientific (and ufological) circles: 'Or all three?'

Some years later, the highly respected Tibetan Buddhist scholar, the late Lama Anagarika Govinda, also saw these lights while staying at Gangtok, barely 30 miles (48 km) from Darjeeling. Govinda, a European by birth, was a guest at the then Maharaja's residence, and recalled in his book *The Way of the White Clouds* how one evening he saw strange lights gliding through the hills. He mentioned them to his host the following morning and was told that the lights had no human origin, and that there was no road in the area Govinda had seen the lights. 'They move over the most difficult ground with an ease and speed no human being could attain, apparently floating in the air,' the Maharaja said, 'Nobody has been able to explain them . . . though the people of my country believe them to be a kind of spirit.' Govinda's host then went on to describe how he had once witnessed the lights moving through his palace grounds towards the place where a temple stood. The site had always been a sanctified place, he stated.

These two accounts are so independent of one another, and the second from a known, highly reliable source, that we can be quite sure of this earth lights lair.

THE CHINA SEA

Writing in *Nature*, 1893, Charles Norcock gave an account of earth lights in the North China Sea, about 17 miles (27 km) south of Quelpart Island near the peninsula of South Korea. On a cold, windy February night, the officer of the watch on board HMS *Caroline* saw strange lights manoeuvering near a 6000-foot (1835-metre) peak on the island. Norcock was called, and his first impression was that they were fires on shore, but as the ship moved 'it became obvious that the lights were not on land, though observed with the mountain behind them'. The lights were spread out in an irregular line and 'resembled Chinese lanterns'. The following night 'globes of fire' were again seen from the ship, this time further north and not near land. Through the telescope, Norcock claimed, the sea could be seen to be illuminated by the lights which 'appeared to be a reddish colour, and to emit smoke'. They would change formation from time to time and were in sight for some hours. On landing in Japan, Norcock read in the newspaper of the appearance of 'the unknown light of Japan', and learned that the phenomenon was known to appear in certain cold weather conditions and was often seen by fishermen in the Shimbara Gulf and Japanese waters. The newspaper article went on to attribute the lights to some kind of electrical effect, and said that they were referred to 'in native school-books'. Norcock was later informed by the captain of the HMS *Leander* that he too had seen the lights in the same vicinity on an earlier occasion. Thinking a ship was in trouble, this captain had approached them, only to find that 'the lights increased their altitude'.

AUSTRALIA

The earth light phenomena of the island continent are given the collective name of 'min min lights'. They are mainly noticed in western Queensland in the area of the towns of Winton, Boulia, Bedourie and Monkira, bordering the Simpson Desert, though they have been reported around Lake Manchester in the east of the state near Brisbane, and they may well appear in other, undocumented locations. They are typical earth lights, appearing as bright globes of light, usually white but sometimes tinged with colour, that flit or float around the landscape, hover, follow people. If approached too closely, they move away or disappear. This happened to one witness, so he gave up the chase. As he drove away, he claimed he suddenly found the light following *him*![10] The light was first noticed by settlers in the vicinity of a former staging post known as the Min Min Hotel between Winton and Boulia, though no one seems to know if the lights were named after the hotel or *vice versa*.

One of the earliest accounts of the lights to be put down in writing by white settlers was that of Henry G. Lamond in 1912. He saw a min min light while on horseback near the Winton-Boulia road. It was winter – cold and windy. At 2 a.m. Lamond saw a 'greenish' light ahead of him. It 'floated rather than travelled' and did not throw a beam like a car head-lamp would – cars were rare in that part of the world at that time in any case. It cast a glow all around it. The man's horse at no time showed any unease. The light approached the witness until he was lit up by the glow it cast. 'Then – phut! – just like that, the light went out,' Lamond stated. He noted no smell or smoke.

In 1953 G. Terris and two companions saw a white, brilliant light over the black-soil plain of the Barkly Tableland. The light faded at intervals and acquired a bright red flash. It moved in circles and spirals above the ground and sometimes shot upwards to an estimated height of 1000 feet (306 metres) only to drop back almost to ground level again. It eventually disappeared by going *into* the ground.

A similar event happened in May 1981, witnessed by Detective Sergeant Lyall Booth. While engaged on stock monitoring near Boulia, he saw a bright light at 11 p.m. one night. He at first thought it was a vehicle head-light, but then realised there was no road in the direction he was looking. The light was below treetop height, and its intensity seemed to fluctuate a little. He kept it under observation for about half an hour, but it did not change its position. Booth fell asleep around midnight, but awoke again at 1 a.m. The light was still visible, but had moved north from its earlier position – if it was the same light. Now it had a slight yellow colouring. Booth was too far away to see any detail, but could make out that the light was between 3 and 6 feet (1 to 2 metres) off the ground, and illuminated the area around it. 'I watched the light for about five or six minutes,' Booth recalled, 'then it suddenly dived towards the ground and went out.' He checked the area the next morning, but could find no track marks where the light had disappeared.[11]

There have been hundreds of such reports over the years, but no one has worked out what the lights are. The aborigines seem to consider the lights as spirits, particularly ancestor spirits or the results of sorcery.

The coastal area of Southern Australia bordering the Great Australian Bight was subject to an outbreak of 'super min mins' in 1988. The most dramatic incident occurred to members of the Knowles family on 20 January, 25 miles (40 km) west of Mundrabilla as they were driving between Perth and Melbourne. A bright light appeared behind them, then paced them at high speed. It swooped down and, connecting with the car roof, lifted the vehicle off the road. The screaming, shouting occupants heard the sound of their voices change and distort as this happened. The lightform dropped the car back onto the road, tearing off the roofrack and bursting a tyre. The Knowles' ran and hid in the bushes until they were sure the thing was not around any more, then repaired the tyre, and, unable to find trace of the suitcase that had been on the roofrack, set off speedily for Mundrabilla. There they met a truck driver who told them, and later the police, that he had been followed in the other direction along the highway by a bright glowing object an hour earlier. Further east, at Ceduna, the family gave a full report to the police. The forensic lab examined black ash which coated the exterior of the Knowles' car, and found it to be burnt rubber. The police took the Knowles' experience seriously, because the pilot of a tuna spotting plane over the Great Australian Bight had also reported seeing a glowing object hovering above a trawler. Moreover, the men from the trawler also claimed in later statements to the police that while the brilliant light hung over them, they were unable to speak properly. Other reports of sightings came in over the same period, describing 'egg-shaped' lights or a lightform like an 'electric light globe'.[12]

HAWAII

In the northern part of the island of Hawaii, around Waimea, there is a localised area where curious lights appear. They float up to 5 or 6 feet in the air, and appear to pulsate. They have the texture of a heavy fog, and some of these who have had close sightings say they make a musical sound. It is believed the lights appear more or less in phase with the new moon.[13]

PERU

Ghost lights are seen in the Peruvian Andes, where they are often referred to as *la loz del dinero* – 'the money lights' – echoing the associations such lights are supposed to have with treasure in other parts of the world. One specific earth lights location is within the complex of curious straight lines on the pampa near Nasca. The highly respected German scholar Maria Reiche, who has studied the Nasca lines *in situ* for over 40 years, has reported moving lights on the desert. There is one point – she will not reveal its exact location for fear of curiosity seekers damaging the fragile linear markings – where a magnetic anomaly affects compass needles. There are also spots, she notes, that the Indians will not walk on as evil spirits live there.[14]

BRAZIL

In Brazil the earth light is known as the *Mae de Ouro* – 'mother of gold'. The tradition is that if one follows the *Mae de Ouro*, the first body of water it crosses will contain gold. Certain locations in Brazil are particularly noted as places where the ball of light appears quite frequently.

When Cynthia Newby Luce moved into her remote, mountain-top home near São Jose do Rio Preto, over 250 miles (400 km) northwest of São Paulo, it was not long before locals asked her discreetly if she had yet seen the *Mae de Ouro*. It transpired that the property was a known location of the light phenomenon, and it had been seen there from beyond living memory. She did eventually see the lightform – in June 1980. It appeared around 7.45 p.m. and she had five other witnesses with her:

> A yellow-orange glowing ball, slightly smaller than a standard volley ball, passed from east to west with the wavering flight of a butterfly about five feet off the ground. It was about 30 feet away from us and was passing between the garage and the house. It made a wide curving turn and headed off towards the stables and vegetable garden near the spring. My gardener, fascinated, went after it. Foolishly he reached out to touch it. The ball faded away to nothing as he put out his hand, then reappeared about fifteen feet ahead of him. He came back . . . unnerved because the phenomenon seemed to him to have intelligence.
>
> I have not personally seen the light again, but my servants have seen it numerous times in the past seven years.[15]

Newby Luce has noted that the colour of the light is reported as yellow-orange as a rule, but occasionally it is seen as a blue-white light. It usually travels from east to west. She also has experienced other unusual happenings in the house. In one case, she happened to notice that a marble-topped antique butcher's table was not present in its usual corner. It was noon on a clear day, and the reason she looked up and noticed the missing item was because of 'a slight shift in the quality of the light'. After 'a few seconds' the table reappeared. On other occasions, there have been sudden, fleeting appearances of people or animals 'like subliminal flashes'.[16]

IRELAND

If earth lights are glimpsed in the Emerald Isle, the tendency, at least in rural parts, is to assign them to fairy activity. In the 1950s Dermot Mac Manus discovered one spot near Castlebar, County Mayo, that has displayed lights at intervals for 'generation to generation'. The place is an earthworked hill known as Crillaun which rises to the north of a small lake or lough. The prehistoric earthworks identify the location as a 'fairy fort' in local lore. Mac Manus interviewed three women who had lived in a farmhouse nearby in their youth. One woman had seen the lights on two occasions. It was late evening and clear weather conditions in both cases, and she 'distinctly' saw multi-coloured lights on Crillaun. They were 'as bright and steady as electric lights would be'. The woman's sister had seen them only once, she told Mac Manus. One evening her attention was drawn to an illumination across the lough, and she saw Crillaun 'brightly lit up with hundreds of white lights'. She saw no colours, but the lights suddenly rose up together and floated over the lough, maintaining their formation, and descended towards another hilltop.[17]

PEAKS OF ILLUMINATION

These Irish hills had a local, pagan sanctity. It will have been noticed in preceding chapters how hills and mountains often figure as earth lights locations, and in Chapter 1 we referred to 'mountain peak discharge' (MPD). It has long been known that certain mountains and ranges produce glows, columns and balls of light. It is even enshrined in an old Japanese haiku:

The earth speaks softly
To the mountain
Which trembles
And lights the sky

The 'Andes light' is particularly well known. It affects the true Andes range as opposed to the coastal cordillera, and appears as a glow at certain points, with occasional outbursts similar to searchlight beams. The light is usually pale yellow. The most brilliant display yet recorded was in August 1906, at the time of a major earthquake, when the whole sky seemed to be 'on fire'.

An artificial MPD effect was created in the 1880s by Selim Lemstrom on a mountain in Lapland. He ranged miles of wire in a spiral formation across the mountain top, furnishing it with brass points and insulators. The wire was earthed over 500 feet (153 metres) down the mountainside. A current was always observed running from the upper wiring to the lower point. When there was strong aurora borealis ('northern lights') activity, dramatic luminous displays would emerge from the wired-up mountain: a column of light would glow vertically from the summit, and a white 'rainbow' effect arched out of the mountainside.[18]

Conversely, natural effects on artificial 'mountains' have been seen too:

the Pyramids at Gizah, for example. The Vice-President of the Institut Egyptien spent a night in the desert near the Pyramids together with a member of the Institut, William Goff. During the middle of the evening, Goff saw a light 'like a small flame' turn slowly round the Third Pyramid. After three circuits, the flame disappeared. Fascinated, the man spent the rest of the evening observing the pyramid. Around 11 p.m. he noticed the same light again, but this time it was a bluish colour. 'It mounted slowly almost in a straight line,' he noted, 'and arrived at a certain height above the Pyramid's summit and then disappeared.' Goff later discovered that the Bedouin had traditions regarding the existence of the light 'stretching back centuries'.[19]

MPD effects have been noted in the Alps, on mountains in Mexico, over the Himalayas, and many other, lesser known peaks. In 1954, for example, the Chief Officer of the SS *Tribulus* reported 'numerous brilliant white flashes' occurring at frequent intervals from certain peaks on the island of Madeira, and visible up to 16 miles (26 km) away.[20] The mountains of Keklujek and Ziaret in the Euphrates Valley have long been claimed to exchange balls of light at sporadic intervals, especially after the long, dry season, with a sound like thunder.[21]

With these moving balls of light, we are clearly approaching areas of phenomena that suggest more complex physics than simple discharge mechanisms. I witnessed such an effect myself, at around 1 a.m. on midsummer morning 1982, while camped with a group of people on the side of the legendary Welsh mountain Cader Idris, which rises above Dolgellau and is not many miles from Barmouth (Plate 30). Idris was a giant, who, it is said, studied the stars. The lore is that anyone who spends the night on its summit will return down the mountain either mad or a poet. The mountain has also been associated with the Hounds of the Otherworld, the *Cwn Annwn*,[22] and there is a tradition that strange lights are seen round its peak on the first night of a new year.[23] Most of the rest of the group had retired to their tents, while two companions and I sat round the embers of a campfire by a tiny lake about halfway up the north side of Cader Idris. Walls of volcanic rock towered above us. A fourth person had his back to us, and the mountain, gazing northwards. Suddenly, my companions started, and we saw a blue-white ball of light hurling itself across the night sky from the direction of the rock wall behind me, but faced by my companions. The fourth witness saw the light pass overhead, and we all saw it disappear over a nearby ridge approximately to our north. I would estimate the light at being one to two feet across. It was intense. Calculating the duration of time it was visible, and the distance it covered to the ridge, the *lowest* speed I could come up with for the light was 600 mph. I recall my chagrin at witnessing such a phenomenon when *Earth Lights* was already in press!

The earth lights phenomenon is literally as old as the hills: small wonder, then, that peaks where such lights appear with some form of regularity have come to be considered holy, or places of spirits or inspiration. If we look around the world, we can select from numerous examples.

One of the classic cases of a sacred earth lights mountain was given by John Blofeld in his *The Wheel of Life* (Rider, 1973) – Wu T'ai, the five-peaked mountain far to the west of Beijing. Blofeld described the flower-carpeted plateau amongst the peaks, nearly 8000 feet (2446 metres) above the North China Plain, with awe. In 1937 he looked on the 300 temples clustered there and declared it was a view which 'might have inspired the original conception of Shangri-la'. It was, he discovered, a holy mountain to the Tibetans, the Mongols and the Chinese alike. During his stay there, he was introduced to the 'Bodhisattva Lights'. Blofeld and a small party of companions visited a temple perched on the southernmost peak of Wu T'ai. It had one tower built on the topmost pinnacle, and its windows 'overlooked mile upon mile of empty space'. It was here that a manifestation of the Bodhisattva was supposed to appear, and, as Blofeld informed me in private correspondence, he understood that the tower had been built specifically to observe the Bodhisattva Lights. Shortly after midnight, a monk roused the party, informing them that the Bodhisattva had appeared. Dressing against the bitter cold, Blofeld and his friends climbed to the tower. Through one of its windows they saw 'innumerable balls of fire' floating majestically past. 'We could not judge their size,' Blofeld remarked, 'for nobody knew how far away they were, but they appeared like the fluffy woollen balls that babies play with seen close up. They seemed to be moving at the stately pace of a large, well-fed fish aimlessly cleaving its way through the water . . . . Fluffy balls of orange-coloured fire, moving through space, unhurried and majestic – truly a fitting manifestation of divinity!' The lights, which apparently always materialised between midnight and 2 a.m., faded from view in the west.

A more modest eminence in Malaya, the 700-foot (214 metres) high hill of Changkat Asah, near the source of the Bernam river and overlooking the village of Tanjong Malim, is also a focus for curious lights. Sir George Maxwell, a government official at Tanjong Malim in 1895, learned that the hill, which rose abruptly from the surrounding plain, was viewed by the villagers as a place where 'every kind of spirit lived'. The mass of stone which formed its highest point was said to be a *bilek hantu* – 'a spirit's room' – because lights could be seen shining from the rock on occasion from the plain below. No one would dare to stay on the hill after dark. A few years before Maxwell's arrival, there had been an attempt to build a trigonometrical observation station on the hill's summit. Foreign workers had to be called in, and then only their leader, Baginda Sutan, was prepared to stay overnight on the hill during the work. The project had remained unfinished, because one morning the workmen found Sutan 'a raving lunatic' – shades, perhaps, of the fate risked by those who sit out the night on top of Cader Idris. Maxwell met Sutan sometime afterwards, and he was still in a state 'of absolute idiocy'. In 1895 another attempt was made to finish the job, with a European this time overseeing the work. This attempt also ran into trouble, due to a roaming tiger at night, making the camp site on the hill dangerous. Maxwell set up a trap for the tiger one night on Changkat Asah, using a goat as bait. As he waited, Maxwell saw two lights far up the Bernam valley moving towards the village. He at first thought they were lanterns being carried along a track, but suddenly and alarmingly, 'in an instant the two lights flew up in the air, and rushed at us . . . two great balls of light sped by within fifty feet of us.' Maxwell and his companions were shocked at the suddenness of this manoeuvre. The lights were 'fiery globes . . . the size of a man's head, and their speed had become terrifying'. He immediately felt them to be some kind of natural phenomenon, but realised he was seeing what the natives felt to be a type of spirit called a *Pennangal* – a spectral, disembodied head. It was doubtless such a manifestation that had turned Sutan's mind, Maxwell concluded. He watched for further lights:

> Soon after several more came drifting down from both sides of the valley towards the river bank, and all, as they reached it, were seized and whirled by the wind in all directions. Before long there were over a hundred to be seen. The wind was fickle and variable, and sometimes a dozen of these balls of light, which were now all around us, would fly down the river together and meet others floating lazily by: they would play round one another . . . .When the wind dropped and there was perfect calm, six or eight would rise, moving in and out among one another as if in some game, and mount up through the air, playing and dancing until they became small bright specks, then slowly sink, revolving and interlacing, until again a breeze would spring up and send them flying helter-skelter up or down the river. We noticed that the lights, as they moved . . . were always round in front and tapered away slightly so as to become somewhat pear-shaped. I image that this shape is caused by the pressure of the air upon the moving body.
>
> . . . All night long the lights beguiled the tedium of our vigil, for they did not disappear unitl . . . the coming of day.[24]

We have here a reference to the 'tadpole shape' we have previously noted with regard to some appearances of earth lights. Maxwell noted that the source of the lights was in part a marshy area, but felt unable to square the nature of the lights with any form of Will-o'-the-Wisp, nor did he feel that St Elmo's Fire, a form of electrical discharge, could be the cause of the phenomenon. The village headman told him that they felt the source of the lights, the spirits, to be the old village burial ground. Maxwell felt sure the luminosities were some form of natural phenomenon, and later associated them with the kind of lights seen at Loch Leven (Chapters 1 and 3).

Mount Athos is a peak over 6000 feet (1835 metres) high at the furthest ip of the promontory which reaches out into the Aegean Sea from the Halkidiki peninsula of Greece. For over a thousand years it has been recognised as a Christian holy mountain, strongly associated with the Blessed Virgin, and there are legends and evidence of earlier pagan appreciation of the sanctity of the place. Through many trials and tribulations, strictly male communities of monks have pursued the holy life on the promonotory. Monasteries survive from the tenth century, but many have been destroyed. Various forms of monasticism are practised on the promontory, the most extreme being the *hesychasts* who lead solitary, ascetic lives and live in rough, unadorned huts perched in impossible positions on cliff faces, or even in caves, seeking union with the Divine Light. Light is a powerful analogy used in the spiritual ambience of Athos. Legend records various kinds of light phenomena there. Around AD 800 it is said that an icon – now known as the Virgin of the Gate – floated miraculously upright in the sea off Athos for seventy years, and then began to emit a column of light that reached far into the sky. It was able to be brought to shore only by one specific holy man.[25] In another legend, the monastery of Simonopetra was founded on its extraordinary, virtually sheer cliff-face, because a bright star or light burning on the rocky ridge was seen around at Christmas time in the fourteenth century. The light persisted in this spot for some days, and it was taken as a sign that a monastery should be built on the spot dedicated to the Nativity. Not long after this, a monk called Dionysiou saw from his cave on Mount Athos 'a fire burning on a great rock' which overhung the water's edge on the southwest shore of the promontory. This light 'flared like a mighty torch' for several consecutive nights. This was likewise taken as a sign that a monastery should be built at the spot.

During his research on the Athos promontory, Andrew Collins found that the monastery of Dionysiou has a small chapel in which can be found the tip of the rock on which the mysterious unfed flame burned. Both chapel and rock are dedicated to St John the Baptist. Collins feels this to be significant as he was able to trace an association between *Sancto Claro* (St Clere) – 'holy lights' – with dedications and festivals connected with St John the Baptist,[26] which supports the earlier observation by Harold Bayley[27] that the midsummer bonfires of the Cornish parish of St Cleer had their origins in the mimicry, or invocation, of the holy light or, in Christian terminology, the Holy Spirit. St John the Baptist is associated biblically with the Holy Spirit, and was the figure used to Christianise the pagan midsummer fires, St John's Day being 24 June.

In his enquiries around Athos, Collins was told by monks that light manifestations do occur from time to time, and these are taken as outward signs of spiritual progress by the monk fortunate enough to perceive them.[28] Sometimes light phenomena reportedly occur on the summit of Mount Athos itself – as in 1821, for instance, when a cross of light was said to have appeared.[29]

A monastery was associated with the Swedish Mount Omberg, on the west coast near Gothenburg, until its destruction by King Gustav Vasa who had become jealous of its power and wealth. The mountain might have got its name from Queen Omma, who is said to have had her castle there before the historic period.[30] There have been reports of light phenomena around Mount Omberg, which rises above flat, fertile country. A young Swedish woman told me of a specific example of this which she experienced while climbing the mountain with a companion. As she looked towards the top, she thought she could perceive soft rays of light beaming out from it. She asked her companion if he could see anything. He looked, and saw the same phenomenon. When they reached the bare, rocky summit, which, she said, has a fringe of trees, the woman claimed that they both could see balls of lights in the branches of the trees. A light would simply 'go out' in one spot, and another suddenly appear elsewhere.

In Britain, the 1800-foot (550 metres) Pendle Hill, an outlier of the Pennines near Clitheroe, Lancashire, undoubtedly had pagan sanctity which was later interpreted as diabolic by Christianity, and so has place names like 'Devil's Footprint' and Devil's Chair' on its 25 square miles (65 square km) of slopes and moors (Plate 24). The area is locally associated with witchcraft and pananormal happenings. There is a legend that the druids worshipped on Pendle Hill, and recently many Iron Age burials have been found on the hill's summit and slopes. The famous Pendle witch trials of the seventeenth century took place in the Forest of Pendle area immediately to the southeast of the hill, in the villages of Roughlee, Blacko, Newchurch and Barley. The Pendle area today is said to have the largest number of practising witch covens in the north of England.

It was upon the Big End of Pendle Hill that George Fox, founder of the Quaker movement, experienced the mystical vision which changed the course of his life in 1652.

There is an enormous amount of folklore concerning local boggarts – the traditional name for ghosts, apparitions and spooklights in the Pendle area. Reports centuries old of 'fairies' disporting themselves like modern troops on the hillside exist, and 'a dwarf-like man' was reported running across the moor for a considerable distance in the late nineteenth century. Boggarts in the form of 'great big dogs, with great glaring eyes' have been allegedly sighted. In early May 1869, at Fence Gate, Forest of Pendle – the traditional meeting place of the witches involved in the 1633 Pendle witch trials – 'something like a fiery flying goose' with a tail like that of a comet was seen flying at a height of over 100 feet (30 metres) by what were said to be credible witnesses. This happened around the time of a local earthquake. This item appeared in the *Burnley Gazette*, and at the end of the article is a plea for some rational explanation to assuage the fears in the local populace occasioned by such a report emanating from 'a place associated with diablerie and supernatural agency from time immemorial'.

More modern 'visions' have occurred around the hill. In 1914 two police officers saw an unusual light in the sky during the early hours of an August morning. It appeared to hang over the Big End of Pendle Hill. The light disappeared after about a minute, but reappeared a short time afterwards, a sequence repeated four or five times. The officers felt they could make out a 'sausage shape' above the light, and this was interpreted as a Zeppelin airship – doubtless because of the time at which the sighting took place. Sixty-three years later, textile worker Brian Grimshaw and a companion saw an almost identical object from Nelson, southeast of Pendle Hill, at 3.10 a.m. on 9 March, 1977.[31] Grimshaw noted that 'there were no rays coming off' the light. The textile workers saw 'grey mist' around the 'cigar-shaped' object – the 1914 policemen saw 'smoke' around their 'sausage' or 'Zeppelin'. In May 1979 a 'blinding white light' was seen to streak over Pendle Hill, and in 1985, on 12 October, a soundless, brilliant white light was seen passing over the hilltop.

Another famous British hill of ancient sanctity is the Tor above the Somerset town of Glastonbury (Plate 31). Archaeology shows that the area was occupied from early Neolithic times. Legend tells that the dramatic, conical Tor, rising 525 feet (160 metres) above the surrounding flat landscape, and capped by the ruins of a church, was the home of the fairy king until his expulsion by St Collen. Glastonbury was a centre of Celtic Christianity and is still a place of Christian pilgrimage as well as a 'New Age' centre. A legend states that Joseph of Arimathea brought the Holy Grail there. Other traditions associate the place with King Arthur. In 1969 I saw, along with many other people, three orange lights half encircle the hill while I was on the summit,[32] and later that year four night-shift workers saw a saucer-shaped object hover over the Tor. A big, fiery-red ball was reported over the hill, also in 1969. In 1970 a police officer witnessed eight egg-shaped 'dark maroon' objects in formation over the Tor. On Midsummer's Day in 1981, witnesses climbing the Tor saw a 'dragon-shaped' glowing orange light writhe out of the ruined church tower on the summit and 'earth' itself near Chalice Well at the foot of the hill. One of the witnesses said 'what we saw was the magnetic field . . . '.[33]

In America there are numerous peaks holy to various Indian tribes. Perhaps the most famous is Mount Shasta in northern California. Light phenomena and curious apparitions have been reported on this mountain from time to time. Shasta was a sacred mountain for the Wintu and other Californian tribes. It is the southernmost peak of the Cascade range, where Arnold saw his flying discs.

We have thus come full circle: we must now try to begin unravelling the revelation Earth's elusive messengers of illumination have so repeatedly been laying before us.

PART 3
A CLOSER LOOK

7 EARTH BORN

It must surely be clear from the information presented in Part 2 that some kind of unexplained phenomenon, conforming to a fairly narrow range of appearances, occurs around the world, and has done so from time immemorial. A remarkable mystery is being quietly displayed – unless one is going to dismiss the range of testimony presented in the foregoing pages as the result of a conspiracy of lies and hoaxes spanning the world and bridging generations. Sooner or later, the lights are going to have to be openly acknowledged. It is already later.

The reader will have noticed repeated characteristics presented by the phenomenon. It often appears near bodies of water – especially lakes and reservoirs, but also rivers and waterfalls. The light produced frequently seems of an unusual nature, very bright, but defined and 'without rays'. There are even tantalising hints in a few reports that the light issues *in only one direction*. Lights are regularly reported near power lines, transmitter towers, mountain peaks, isolated buildings, roads and railway lines. The lights like to haunt quarries, rocky ridges, mines and caves. Some witnesses have reported air pressure and temperature changes in the vicinity of the lightforms. Outbreaks of earth lights are sometimes accompanied by columns of gaseous material that are often traditionally interpreted as 'white lady' or 'black monk' ghosts, usually dependent on whether they are seen at night or in the day. (A modern version of this is probably the 'silver-suited alien'.) Observers who come close to the lightforms quite often report a 'buzzing in the ears', or humming or whistling sounds. The lights are highly manoeuverable. Fiery coloured, amber, yellow or white globes of light are the norm, but colour changing and shape-shifting are common. When white lights are closely observed a red centre is frequently remarked upon, and sometimes a teeming inner activity is seen. Descriptions of the lights as being like 'Chinese lanterns' and the size of 'basketballs' recur frequently. Some reports suggest the lights display the effects of mass at times, yet, at others, to behave as if weightless (the light witnessed by David Kurbin at Pinnacles National Monument, California, is a particularly interesting case in point). It seems as if the lights are on the very edge of physical manifestation. This remarkable characteristic may explain why some aerial lights show up on radar while others do not. It may be recalled that in Hessdalen a light observed visually only registered intermittently on Project Hessdalen's radar screen.

Whatever the lightforms are, they tentatively intrude into our time and space, into the material world, and thus have to use energy to make an appearance, and to move. Their overriding characteristic is that, whether for a long period or a short, intensive outbreak, the lights *relate to a given locale or region*. This is the crucial clue. Whether they are bizarre lifeforms, the produce of exotic physics, or craft from outer space or other dimensions, there has to be something such terrain-related lights need in these geographical zones at certain times. That requirement can only be some source of energy that can facilitate their appearance and mobility, if not actual existence. But what can this be?

As indicated in Part 1, the obvious place to look is in the nature of the Earth's crust itself – the geology – and the energies released in seismic stress and strain. In Part 1 we discussed how these forces can, during earthquake conditions, produce localised points of thousands of volts per metre at ground level – and even in the eighteenth century the researcher von Humboldt was able to measure changes in atmospheric electricity at the time of earthquakes. Certainly the behaviour of earth lights in appearing near sharp projections in the landscape, or following possible conductive material in roads and railway lines (and, perhaps, being attracted to metallic vehicles), as listed above, suggests an electromagnetic component in their make-up.

Bodies of water, too, would provide relatively subtle sources of geological strain: lakes and reservoirs are known to cause settling and pressures in the rock strata and faulting beneath and around them, triggering numerous tiny 'microquakes'. Only three earthquakes were recorded by distant seismographic stations between 1960 and 1963 around Longarone, Italy, for example, but when a reservoir there was filled as many as 400 low-magnitude tremors were registered at the dam site. This activity was blamed for a disaster which happened there in 1963. Before the building of the Monteynard Dam in France in 1963, no earthquakes had been recorded in the area, but between the reservoir's first accumulation of water and 1970, 64 earthquakes took place there. The strongest ones always happened when the greatest volumes of water were present. A dam in India was built in a seismically inert region, but over a seven-year period after the dam's construction, thousands of little quakes were registered. So well known is this association between bodies of water and local seismic activity, that reservoirs in several countries are now being seismologically monitored at the recommendation of UNESCO. Persinger points out that accumulations of water contribute significantly to the resistivity of underground minerals, a factor that could increase the likelihood of light displays in an area. Also, moving water such as rivers and waterfalls, not to mention the sea crashing and heaving against coastal rocks, produce ionisation within the local atmosphere which would probably contribute supportive electrical conditions for earth lights, as might the electromagnetic fields around power lines, other circumstances prevailing. Higher ionisation levels can also occur in the mouths of caves, and it has been suggested this accounts for an alleged higher incidence of lightning strikes at cave entrances.

The reports of subterranean sounds at some earth lights locations (Hessdalen, Yakima reservation, and Hafren Forest area, for examples) also clearly support some kind of seismic association with the appearance of lights. References to aurorae – 'northern lights' – appearances, or flashes in the sky over such zones (Hafren, Hessdalen, and Yakima again, the Uintah Basin, and so on) suggest unusual atmospheric electricity. The appearance of odd cloud formations (well described at Yakima) have also been observed in association with earthquakes.

The evidence thus indicates the presence of geological and atmospheric conditions conducive to the appearance of light phenomena, conditions occasionally existing in electrical storms (producing 'ball lightning') and perhaps a little more frequently in earthquakes (producing 'earthquake lights' and 'earthquake lightning'). However, most cases of earth lights do not seem to be associated directly with either circumstance, and most thunderstorms and earthquakes do not produce unusual light phenomena. It seems that a delicate balance of conditions have to come together for lights to appear. Perhaps materials in the Earth's crust undergo temporary change due to pressure or other stimuli, allowing the transient formation of superconductors, for example. The matrix for earth lights appearances is certain to be complex, and it will take a long time to tease out all the factors that may be involved, and some of these may have interchangeable roles and not always be present at any given occurrence of light phenomena.

The tectonic strain theory suggests that *low magnitude* seismic disturbance may be sufficient to provide the conditions for light phenomena in certain circumstances. If it is stress and strain in rocks that produce the electrical or electromagnetic conditions, then the build up of pressure is the chief factor, whether the release is a relaxed, unnoticed process or a violent readjustment causing an earthquake. The pressure changes within a local portion of the Earth's crust may be occasioned by many different causes: seismic disturbance close at hand or at a distance (if such a source vibrates the local geology or massages local faulting); periodic flexing due to the effect of lunar, solar and possible planetary gravity effects; the movement and settling of local rocks due to bodies of water, collapse of subterranean caverns or mine workings. Ambient electrical conditions would be modified by these sort of movements, as well as by extra-terrestrial effects on the ionosphere – such as charged particles from solar flares.

So earthquakes do not have to happen in earth lights zones. Nevertheless, we would reasonably expect to see in such areas up to three clues suggesting the possibility of a geologically based energy source being available at them: the presence of local surface faulting, the presence of mineralisation, and in many cases (though not necessarily all) a history of seismic activity in the general region. Whether such zones exhibit all these primary clues depends in the first instance on whether appropriate, detailed study has been carried out, and this is often, even usually, not the case. Even if a zone is studied, all three of the clues may not be present. However, if faulting is present, it is probable that mineralisation will be also, as mineral enhancement tends to occur along fault margins. This is why there are often mines and quarries in faulted areas – and both these types of locations, as well as caves and bare rocky ridges, are typical haunts for earth lights. (The Japanese researcher Yasui has noted that EQLs, in general, are restricted to certain areas in an earthquake, none of them the epicentre. Perferred locations seem to be mountain summits in quartz-diorite faulted rock.)

The probable reason for the importance of ore bodies in an earth lights zone is that when minerals are subject to various kinds of pressures or strains, the electrical fields around them can change. John Derr and Michael Persinger elaborate on the possible importance of minerals in a paper in *Experientia* (42, 1986):

> Another potential mechanism involving the semiconducting properties of polymetallic ore bodies has been proposed by Demin *et al.* Their idea involves electrical discharges from cracks which are amplified by unusual occurrences of semiconducting minerals. These minerals would then become transistor amplifiers or thyristors in the bodies, with high-frequency stress waves producing piezoelectric polarised p and n junctions. This theory entails the generation of ultrasonic waves and electron emission, in addition to luminescence, and suggests that light phenomena might be associated with polymetallic ore bodies near the surface.

Repeated references in Part 2 to minerals in areas of earth lights activity may have been noticed by the reader – it is a connection which will be reinforced in this chapter. We can also recall the fact that medieval miners of tin and copper actually looked for the emission of lights from the ground as a guide in their search for new mineral veins (Chapter 2).

A local seismic history will not necessarily be present at every earth lights location, because faulting will not necessarily be associated with strong, recorded movement in recent history. Forces may wax and wane in an area for a long time before a quake sufficiently large to register on existing monitoring networks actually occurs. As we shall see in Sweden, for example, such a place may produce lights only during the pre-quake period, and stop producing them when a quake does occur, releasing the all-important stress. The implication of this is that geology resistant to sudden movement might actually augment the appearance of lights. It is the effects, and perhaps also something about the nature, of geological *stress fields*, that seem to be the major catalyst in the production of the lights.

When Michael Persinger or John Derr study an earth lights area, they look for the record of minor or mid-magnitude quakes and try to associate them statistically in time with the reported incidence of light phenomena. As we shall see, they have been successful in this to a large extent. But it is also suggested that some conditions are created by pressures that are below available recorded levels in certain areas. Even low-level stress in a landscape can add up to enormous amounts of energy, and individual instances of how that energy is channelled, transmitted and focused within local geology may result in phenomena equal to that produced by more robust stimuli. Even within the range of measurable earthquakes themselves, EQLs are sometimes seen in association with small events and yet are missing from much larger ones.

The examples of earth light locations described in Part 2 have been assembled from data that became available to me from many disparate sources. I did not pick and choose the geographical areas myself, any more than have Persinger, Derr or any other researchers who may be said to be interested in the same general avenues of enquiry. Consequently, some of the sites have been investigated from an earth lights perspective in some depth, others only tentatively, and some not at all. The uneven range of evidence this situation creates cannot be avoided until such time as more resources are put into this kind of research. But in this chapter I will trawl through those sites mentioned in Part 2 for which some level of information, detailed or tentative, is available. It will prove sufficient, I feel, to convince anyone without their own axe to grind that the association of geological factors with earth light appearance is reasonable and probable, thus providing the single most powerful clue we have with regard to so-called UFO phenomena.

BRITAIN

I consider here earth lights sites mentioned in Chapter 3.

SCOTLAND

I am not aware of any specialist work having been undertaken by anyone on the geological nature of the lochs which have been reported as prone to earth light activity. But we can here make some observations which, while basic, leave little in doubt. Loch Ness, scene of sporadic light phenomena, including a 'Chinese lantern', lies as we noted on the huge Great Glen Fault, possibly Britain's premier fault line. Low-level seismic activity is fairly constant around this feature, though, thus allowing a regular release of stress: this could explain why light phenomena does not have a greater incidence here.

Of the other earth lights lochs, Loch Tay, where 'balls of fire' were seen skimming the water surface, is crossed by a major fault running SW–NE. Loch Maree on the west coast has a major fault running along its northernmost shore, and it passes through Taagen where curious 'flames' were often seen, and treated with superstitious dread. The bay of Ob Mheallaidh at Upper Loch Torridon, which provided sanctuary for an 'often seen' earth light, is 2 miles (3 km) from where major faulting meets the loch. In addition, as local man Dr J. H. Fidler points out, the bay 'is situated over Lewisian Gneiss . . . this rock is full of faults . . . ' (personal communication). Loch Rannoch, well known at one time for its *gealbhan* or ball of fire, is traversed by two major faults. Loch Campbeltown, scene of light phenomena, is bounded on its western and southern sides by major faulting. Loch Leven, most famous of the earth lights lochs, has its western entrance on Loch Linnhe, directly on the line of the Great Glen Fault. Loch Carloway, on the Isle of Lewis, which had its own light, has an important local fault pass under its southwest corner.

So, not considering Loch Ness, seven of the eight Scottish lochs known to have generated earth light reports are associated with *major* faulting, quite apart from local fissures, and the eighth is on a locally important fault. Many, if not all, of these places are seismically active, too. As a random check, I scanned earthquake epicentre records of the British Geological Survey (BGS) for the periods 1967–78 and 1982–4. Much activity has occurred at Loch Leven in these years – perhaps such strain-release may account for the modern drop in lights incidence. There have been epicentres at Loch Maree, Loch Campbeltown, Upper Loch Torridon (directly around the Ob Mheallaidh location), and adjacent to Loch Rannoch and Loch Tay. In other words, we have evidence of tectonic strain, particularly at low levels, around these places, in addition to the presence of faulting alone.

Gruinard Bay, where a light 'without rays' used to drift from the area of Coast village to the rock of Fraoch Eilean Mor, does not display major faulting, but the very part of the bay the light traversed is over gneiss, indicating probable local faulting. The BGS records mentioned above, show epicentre activity in the bay. The isle of Luing in the Firth of Lorne lies close to the line of the Great Glen Fault, and has bands of intrusive rock cutting completely across it at a number of locations, so local faulting can be assumed, and there has been adjacent epicentre activity in recent decades. Benbecula, another island associated with light phenomena, lies

Table 1 The spectrum of earthquake effects covered by the modified Mercalli scale of quake intensity.

I	Not felt. Marginal and long-term effects of large earthquakes.
II	Felt by persons at rest, on upper floors, or favourably placed.
III	Felt indoors. Hanging objects swing. Vibration like passing of large trucks. May not be recognised as an earthquake.
IV	Hanging objects swing. Vibration like passing of heavy trucks, or sensation like a jolt, like a heavy ball striking the walls. Standing motor cars rock. Windows, dishes, doors rattle. Glasses clink. Crockery clashes.
V	Felt outdoors; direction estimated. Sleepers wakened. Liquids disturbed, some spilled. Small unstable objects displaced or upset. Doors swing, close, open. Pictures move. Pendulum clocks stop, start, change rate.
VI	Felt by all. Many frightened and run outdoors. Persons walk unsteadily. Windows, dishes, glassware broken. Books off shelves, pictures off walls, furniture overturned. Weak plaster and masonry cracked. Small bells ring. Trees shake visibly.
VII	Difficult to stand. Noticed by drivers of motor cars. Hanging objects quiver. Furniture broken. Damage to masonry, including cracks. Weak chimneys broken at roof line. Fall of plaster, loose bricks and stones. Waves on ponds, water turbid with mud. Large bells ring.
VIII	Steering of motor cars affected. Partial collapse of walls. Twisting and fall of chimneys, factory stacks. Frame houses moved on foundations if not bolted down, loose panel walls thrown out. Branches broken from trees. Changes in flow and temperature of springs and wells. Cracks on wet ground and steep slopes.
IX	General panic. Much masonry heavily damaged. General damage to foundations. Serious damage to reservoirs. Underground pipes broken. Conspicuous cracks in ground. In alluviated areas sand and mud ejected, earthquake fountains, sand craters.
X	Most masonry and foundations destroyed. Serious damage to dams and embankments. Large landslides. Water thrown on banks of canals, rivers, etc. Rails bent slightly.
XI	Rails bent greatly. Underground pipelines completely out of service.
XII	Damage nearly total. Large rock masses displaced. Lines of sight and level distorted. Objects thrown in the air.

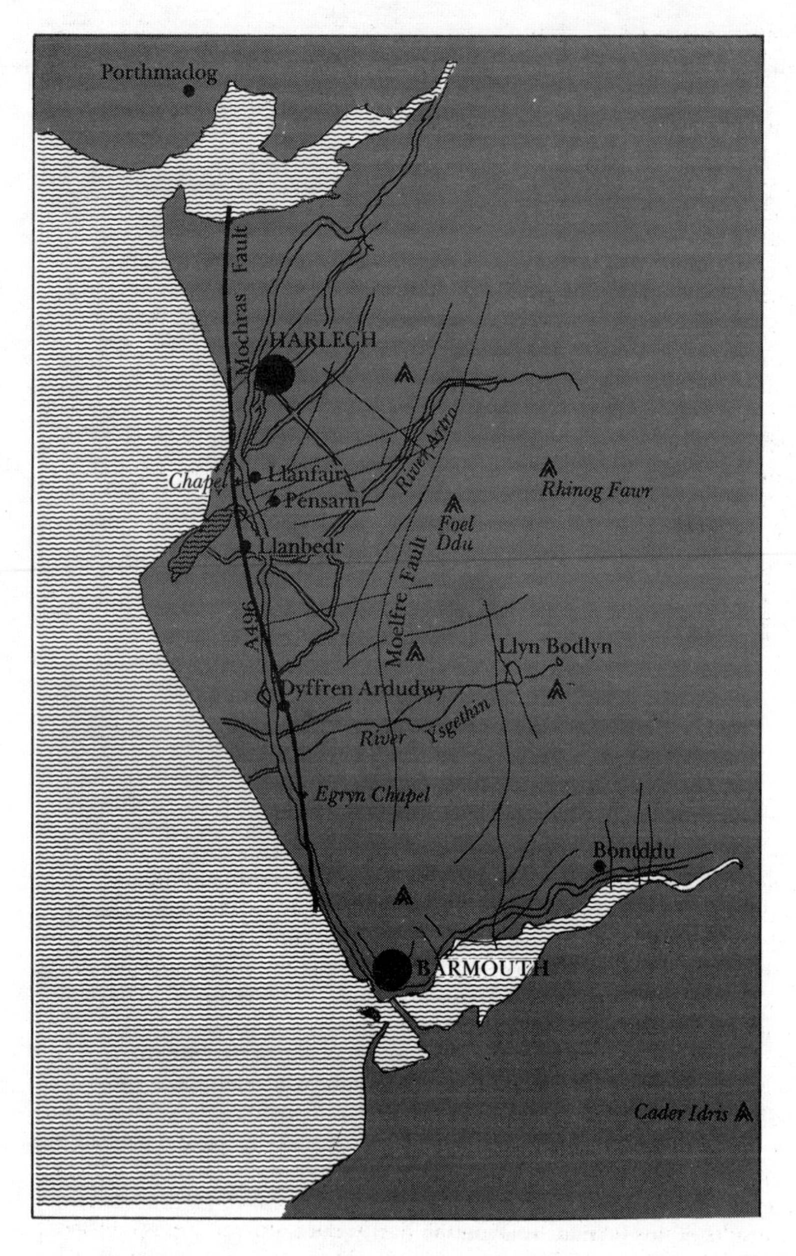

FIGURE 13 Faulting in the Barmouth-Harlech area. Lights in the 1904–5 wave seemed to relate particularly to the important Mochras Fault (see Table 2).

about 11 miles (18 km) west of the great Minch Fault, and a major thrust runs through the east side of the island. As Benbecula is largely composed of Lewisian Gneiss, local faulting can also probably be assumed.

In Chapter 3 examples were given of lights being seen around the Oban-Benderloch area. This region is bounded by a major fault on the east, and has local faulting. Moreover, it is a part of Scotland that has a history of seismicity (maximum intensity 5–5.5 Modified Mercalli - see Table 1 - between 1800 and 1970).[1]

On the eastern side of Scotland we noted the light seen near the coast by the people of Latheronwheel, south of Wick. The light appeared very close to, if not actually on, the course of the major Helmsdale Fault. Finally, Gamrie Bay on the north Aberdeenshire coast: this has major faulting 2 to 4 miles (3–6 km) either side of it, and occurs on a boundary between Old Red Sandstone and Upper Dalradian slates, mica and schists, so local faulting is likely. The boundary, in fact, seems to be just about where the Gamrie light was usually reported, around the old St John's church on the bay's western perimeter.

Thus all the Scottish earth lights locations that came to hand during research for this book can be placed in areas of noteworthy geological significance.

I have no geological information on the sites in Northern Ireland, so we can move directly to Wales.

WALES

One of the areas that has had the most thorough earth light study is the Barmouth–Harlech region. When *Earth Lights* was written in 1981, we had no detailed knowledge of the faulting or seismicity in the area. The new geological survey of the district was not then available, and we were unable to obtain a special copy as there was a strike that lasted long enough to thwart the book's schedule! Although we knew there would be faulting in the region, we had no idea how remarkably the pattern of locatable light events would mirror the geology. The crucial geological feature in the area, I have subsequently found out, is the Mochras Fault - a deep-rooted (1995-foot - 610-metre - downthrow) feature that virtually links Barmouth and Harlech (Figure 13). Most of the geographically positionable light events were strung out along it like beads on a thread. Many field visits have been made to the area, so precise locations have been confirmed. Initial results of this further work on the Barmouth outbreak were published in *New Scientist*,[2] but the histogram presented here (Figure 14) represents a further development. In all, we have to date positioned 21 sightings (Table

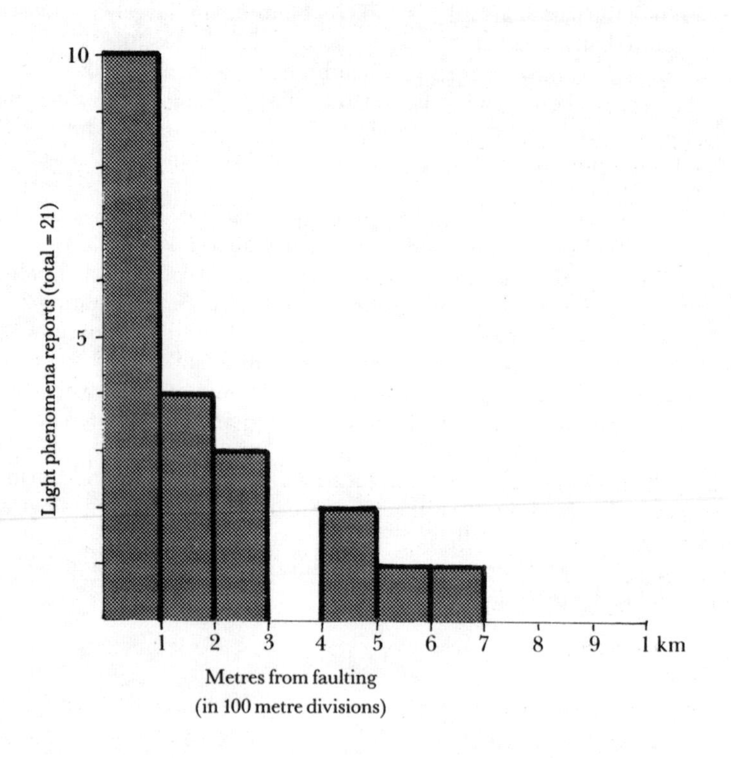

FIGURE 14 Histogram showing the distance of light phenomena from surface faulting, as reported in the 1904–5 events around Barmouth. Only events that could be geographically positioned from the reports could be used for this study. The horizontal axis shows distance from faulting in 100-metre increments. It can be seen that there were dramatically more reports of lights closer to faulting.

2): there were many more, and all of them likely to be close to faulting, but descriptions of their locations have proved too vague to use without risk of criticism. In a few cases in Table 2 there is a doubt of a few hundred metres as to precise location, and where this has happened I have erred on the side of caution for the histogram. Even with this conservative approach, we can see from the histogram that not only did most of the reported and locatable light events occur close to recorded surface faulting (usually but not solely the Mochras Fault), but also that *incidence increased with proximity to faulting.*

Two key locations were the isolated chapels of Egryn and Llanfair. Egryn

Table 2 Barmouth area light phenomena.	
Four separate light phenomena events in fields and roadway around Llanfair Chapel	0–20 m from Mochras 0–20 m other faulting
Strange lights 'coming together with a loud clap' at Pensarn	150–250 m from a fault; 500–600 m from Mochras
Light on hillside above Egryn Chapel	400–500 m from Mochras
'Three brilliant rays' on road near Egryn Chapel	0–150 m from Mochras
'A kernel of light' above road north of Barmouth	400–500 m from a fault
Intense yellow light at Egryn Chapel	0–100 m from Mochras
Two flickering lights at Egryn Chapel	0–100 m from Mochras
Deep yellow light near Egryn - Barmouth road	150–250 m from Mochras
A 'bright light' on road in front of Egryn Chapel	0–100 m from Mochras
Light over road east of Barmouth	400–500 m from a fault
Moving, protean lightforms at and near Llanbedr; passed crossroads	200–300 m from a fault
Blood-red light just above road near Egryn Chapel	0–400 m from Mochras
Glittering light above Egryn Chapel roof	0–100 m from Mochras
Moving light east of A496 road	500–600 m from Mochras
Several lights on hillside east of Egryn Chapel	400–500 m from Mochras
Light on road in front of Egryn Chapel	0–100 m from Mochras
Three 'columns of fire' on road at Dyffren	100–200 m from Mochras
Bar of light over road near Egryn Chapel	0–100 m from Mochras

Chapel is less than 100 feet (90 metres) from the Mochras Fault, while Llanfair Chapel is situated precisely on it. The lights seen coming out of the ground near Llanfair Chapel described in Chapter 3 were definitely emerging directly from the fault – there is simply no doubt about that.

But the 'Egryn Lights' occurred significantly in time, as well as in space. The historical unit of the BGS in Edinburgh kindly supplied me with a history of seismic activity in Wales over the turn of the century. It is now clear that Wales, and particularly northwest Wales where the Barmouth or

Egryn lights occurred, was experiencing unusual seismic activity over a period of about 14 years, the Egryn lights occurring in the latter part of this period. Rocks and faulting were definitely undergoing the wax and wane of stress. In 1892, Pembroke, in southwest Wales (scene of the 1977 flap), experienced a quake and a 'coincidental' meteor. In 1893, Carmarthen had a quake. On 19 June 1903 there was a quake centred around Caernarvon (about 25 miles - 40 km - north of the Egryn area), with subsequent after-shocks. On 21 October 1904 there was an earthquake at Beddgellert, only 18 miles (29 km) from the Egryn area. (The outbreak of lights between Barmouth and Harlech commenced no later than December 1904). This cycle of seismicity came to a close with a big quake at Swansea in June 1906 - only a few months after the great San Francisco earthquake, as it happens. There are also accounts of quakes (and light phenomena) going back centuries in northwest Wales. And in modern times, in July 1984 to be precise, Britain experienced its largest (5.5 on the Richter scale) earthquake in a century: this was centred in the general area of Porthmadog on the Lleyn peninsula, only 5 miles (8 km) north of Harlech.

When I gave an account of some of this work - excluding the seismic history which was not in my possession at that time - at the 1983 BUFORA International Congress in England, I was energetically criticised by American UFO researcher and scientist Stanton Friedman. (I was also bitterly attacked at the same conference by what the *Guardian* called 'a pinkly irate maiden' in the auditorium who felt earth lights were of no consequence and only UFOs mattered. . . .) Friedman has a great belief in the 'nuts and bolts' reality of alien spacecraft - he has a background in rocket engineering - and from his attitude it appreared he felt that the thesis I was presenting somehow threatened his convictions. His main point was that a road connects Barmouth with Harlech, and the apparent distribution of the lights was an artefact created by this (people used the road, so most reports would come from that location giving a false sense of the lights' distribution). It is true that a road (the A496) runs between Barmouth and Harlech, but it is also the case that that it weaves along the Mochras Fault. The view from the road extends many miles to the west over the coastal strip and into Cardigan Bay, and a mile or more eastwards to the first crest of the hills forming the area of ancient rocks known in geology as the 'Harlech Dome'. It is difficult to see how the marked trend of increased light event incidence with a proximity to faulting measured in metres would be created simply by phenomena being observed from this roadway, if it did not also approximate the course of an important fault - lights could be seen miles away!

It is in any event the case that not all the witnesses were on the A496 - some were on minor roads to east and west of it, and at right angles to it. In the light of the expanded material I have presented above it seems clear to me, at least, that the distribution of the lights relates more surely to the presence of the fault than to the road. Moreover, when this case is placed in conjunction with the other evidence presented in this chapter, this becomes even more obviously the situation.

Before leaving this area, I should mention that it has some mineralisation in the forms of nickel, iron, copper, gold and zirconium amongst others. Zirconium is possibly especially interesting as during the writing of this book I have been contacted by a university researcher looking into the pyroelectrical properties of zircon. It seems this mineral can produce electrical charge when heated, and the researcher sees it as a possible candidate for the production of light phenomena.

Forty miles (64 km) inland from Barmouth, the place where the vicars saw lights emerging from the ground during their 1905 skywatch close to Vroncysyllte, Llangollen, has shown itself to be riddled with local faulting. Indeed, there is an actual 'Aqueduct Fault' there, named after the location where the clergymen stood!

A study of the 1977 outbreak around St Bride's Bay in the county of Dyfed, southwest Wales, has provided similar results to those obtained in the Barmouth area study. This part of Wales is geologically famous – a frequent target of school and college geological field trips. Different rock types of different ages cluster here because the landscape has experienced significant faulting and folding during both the Caledonian and Armorican orogenies (mountain-building periods) hundreds of millions of years ago. One can stand on the beach of St Bride's Bay and see faults exposed in the surrounding cliffs. It is a classic background for an outbreak of light phenomena. Again, it was possible to place fairly accurately a certain number of the reported sightings, a task undertaken by Paul McCartney. (Detailed geological maps of the northern part of the region affected by the wave of phenomena in 1977 were not available when the work was undertaken.) In the end, McCartney had 26 locatable sighting positions for that part of the region for which he had sufficiently detailed geological information. On that basis he produced a histogram, a version of which is given in Figure 15. Like the Barmouth histogram, the horizontal axis shows distance from faulting in 100-metre (110-yard) increments. It can be seen how similar the Dyfed relationship of sightings to recorded surface faulting was to that around Barmouth: a much increased incidence with proximity to faulting.

In the Dyfed wave, witnesses were situated in many different locations – roads, fields, houses, beaches, clifftops, school playgrounds and so on – yet the pattern related to faulting still holds, suggesting that the distribution of the Barmouth–Harlech lights was related to the Mochras Fault rather than being some artefact caused by the presence of the A496 road which happened to run beside it.

In Chapter 3 a window area extending across North Wales was indicated, where UFO activity has been reported periodically, with

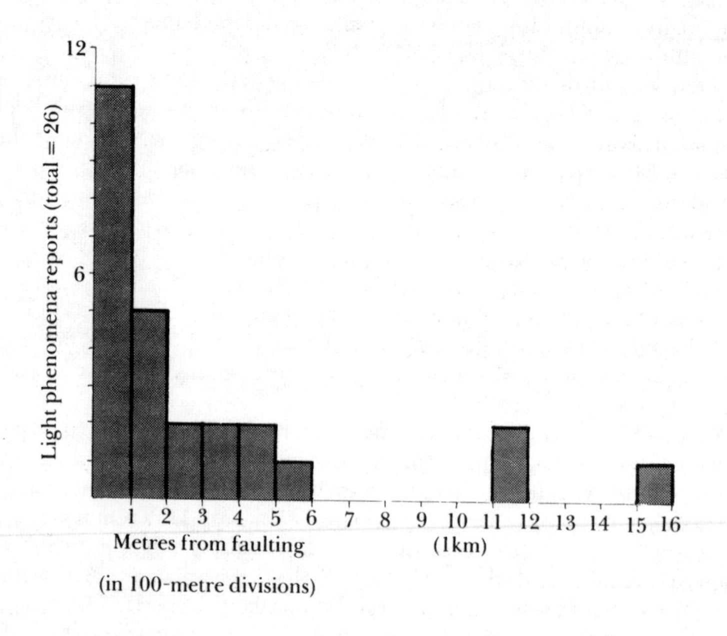

FIGURE 15 Histogram showing distance from faulting of reported UFOs in the 1977 Dyfed wave. As in Figure 14, only geographically locatable reported events could be used, distance from faulting is indicated in 100-metre divisions, and the increase of reported sightings with proximity to surface faulting is strongly marked.

increases of activity in 1984 and 1986. Paul McCartney has studied the area and the locations of the reported sightings from a geological standpoint. The whole area has been actively faulted and folded on more than one occasion, but the disturbed Silurian beds (of 400 million or more years ago) in the region seem to provide particularly well correlated geological features for the reported light phenomena. An earlier case in the area that was quoted was the remarkable incident at Brookhouse Mill, close to Denbigh, in 1979. Here the Denbigh Fault, a very prominent north–south feature, separates highly faulted outcrops of rock. Brookhouse lies on what geologists refer to as Lower mottled Permo-Triassic Sandstone. Significantly, the Llanrhaiadr Fault pushes into this rock formation only at Brookhouse and its neighbouring community of Kilford. This location thus has the highest geological qualifications for staging an earth lights show.

The main wave in the area in 1984 commenced with the dramatic happenings around Llangernyw, when investigator Margaret Fry was awoken by the commotion caused by the UFOs the night before – 14 April. The village is adjacent to the West Llangeriew Fault, and Cwm Canol, Gwytherin, one of the landing sites, is directly next to the East Llangeriew Fault on disturbed Silurian sediments. Moreover, on 15 April a Welsh earthquake occurred, centred near Felindre, a small village about 55 miles (88 km) southeast of Langernyw. This was one of Britains's strongest recent quakes, registering 3.3 on the Richter scale and affecting a 400-square-mile area. It is interesting to note that the main wave of British UFO activity for 1984 occurred in the two weeks preceding the quake.

Prior to the Llangernyw area outbreak, lights and aeroforms had been seen off the North Wales coast over Colwyn Bay and Abergele. Colwyn Bay is on disturbed and faulted Silurian sediments and Abergele is sandwiched by the Siamber Wren and the Kinmel Park Faults. It is significant that both these faults extend seawards, which would further enhance the chances of sightings.

Phenomena were reported at various times in July 1984 at Llanfair Talhaiarn, 4 miles (6 km) from Llangernyw. This location is surrounded by highly faulted, disturbed Silurian sediments, and July was the month of Britain's already-mentioned 5.5 Richter scale quake centred on the Lleyn Peninsula about 25 miles (40 km) to the west.

Another wave occurred over late 1985–early 1986. 'Yellow eggs' of light were seen rising from the moors west of Bylchau. This area is also on disturbed Silurian rock, and is severed by the Bylchau Fault. During this wave, Meliden, near Prestatyn on the coast, formed one of the key foci for lights. This lies on coal measures and there are several prominent faults with north–south trends very close, including the Prestatyn Fault. It is on the edge of a zone of very pronounced tectonic disturbance. Shotton, another key location during this period, similarly lies on carboniferous coal measures and has prominent faulting from the northwest, where Connah's Quay is situated and phenomena were reported.

All the other locations mentioned in this region in Chapter 3 fall on faulted or disturbed geology and it is clear that the North Wales window possesses a rich collection of the geological features and seismological history one ideally associates with earth light zones.

Geological information on the Hafren Forest–Elan Valley areas is scarce. A field study of part of the Elan Valley was conducted in the late 1890s (H. Cecil Moore, Woolhope Club Transactions 1896–7) prior to the forming of all the reservoirs, and faulting was noted around the Caban Cock dam (very close to one of the reported light phenomena locations) and extending up the valley. But we can be sure the Elan and Hafren areas in general are subject to tectonic stress because on 15 April, 1984 the 3.3 Richter scale quake (referred to above) occurred, centred around Felindre, a village only 19 miles (30 km) to the east. It is interesting to note that the Earth Mysteries writer Janet Bord, who at that time lived in the area, noted an inexplicable light in the sky on the night of the quake. The curious lightform in the disused Dylife mine was seen in 1984, too – but I do not have the precise date of the event.

ENGLAND

Paul McCartney carried out a detailed study of the relationship between faulting and the distribution of stone circles in England and Wales for *Earth Lights*. He found that all stone circles in that designated region were within a mile of a surface fault or lay on an associated intrusion. Less detailed preliminary study indicated a similar pattern in Scotland, and it is known from the work of Belgian researcher Pierre Méreaux[2] that the great megalithic complex in Brittany around Carnac also lies on granite intrusions, is surrounded by faulting and has the most intense record of seismicity in France. An *intrusion* is a body of igneous rock, such as granite, crystallised out of the magma beneath the Earth's crust at times of tectonic upheaval and intruded into or emplaced within the crustal rocks at times of tectonic upheavals hundreds of millions of years ago. (Extrusive rocks have been crystallised from magma at the Earth's surface by volcanic action.) Where intrusions have occurred, there has to be faulting on a number of scales. So there is no difficulty in making geological associations with light phenomena that appear around stone circles. Castelrigg, where T. Sington saw large white balls of light, lies on a granitic intrusion, and therefore considerable minor faulting can be assumed; more substantial recorded faulting occurs about 3 miles (5 km) from the site.

Granite is one of the more radioactive rocks, and I have a feeling that certain levels of radiation may also be a contributory factor in some earth lights appearances. For instance, within certain prehistoric stone structures, such as dolmens (upright stones with a large horizontal stone capping them) and fogous or souterraines (subterranean stone chambers and passages), small-scale light phenomena have now been reliably observed. These particular stone structures are made of granite and their interior atmospheres are considerably more radioactive than normal background. This matter is dealt with at greater length in *Places of Power*.[3]

The hamlet of Linley, where balls of light and vaporous columns were seen in 1913, displays no dramatic geological significance at first glance. But it is none the less noticeable that surface faults do encompass the hamlet like callipers less than a mile (1.6 km) to both east and west, and a drift boundary crosses between St Leonard's church and Linley Hall, which may possibly indicate highly localised geological disturbance.

In Chapter 3 the account of the remarkable outbreak of light phenomena around Burton Dassett in south Warwickshire in 1923 and 1924 was given. One of the principal investigators of this case, David Clarke, records:

> The lights centred on the pre-Norman All Saints Burton Dassett church. The hills are riddled with minor faults and the church is situated over the actual 'Burton Dassett Fault'. It is built into the hillside in a series of step-levels, and there are a number of carvings of dragons/serpents on the supporting pillars and nave inside.[4]

Truly a dragon's lair. The church provides a vertical projection on a fault much like the lonely chapels on the Mochras Fault. A seismic association is further underlined by the sudden, brief reappearance of the Burton Dassett light in January 1924. As Clarke further comments: 'It is most interesting that a fierce earth-tremor hit Herefordshire on the night of Friday, January 25 1924, the same night as the "ghost" light reappeared on the Burton Hills, this time casting an even more luminous brilliance.' The *Leamington Chronicle* made this same association, remarkably, at the time: 'Simultaneous with the reappearance of the "ghost" of South Warwickshire comes the report of an eath tremor in the West of England.'[5] The earthquake was also accompanied by 'a weird arch of auroral light which lit up the sky for miles around' in the north of Scotland.[6]

Dartmoor is a huge area of granitic intrusion, with high background radiation levels and numerous old tin mine workings. It is not surprising, therefore, that our survey of British earth lights cases in Chapter 3 concludes with examples of lights rising, falling and moving over this landscape. Other light events doubtless occur unobserved on Dartmoor's wilder, uninhabited tracts.

Chapter 4 was devoted to an account of the findings of Project Pennine. A full correlation of the lights with geology and seismicity has not yet been completed, but Clarke, Roberts and colleagues have already been able to note some telling points of interest.

The Longdendale area (the Project's 'Area 1') not only has a string of reservoirs overlying a line of faulting in the valley, but the high ground on either side is considerably faulted. The infamous Devil's Elbow on the B6105, scene of the eerie black shape encountered by John Davies, is crossed by a fault, and the extraordinary light that lit up the valley seen by Barbara Drabble in 1970 stretched from Bleaklow to Black Hill, approximating the course of faults that cross Longdendale. Figure 16 shows most of the other locatable light phenomena referred to in the text in relation to the faulting of the area. None of the locations are more than hundreds of metres from primary local faulting.

Remote Wessenden Moor (also in Area 1) has numerous faults running through it, and contains reservoirs. A fault crosses the isolated A635 road which traverses the moor at a point where some of the close sightings of light phenomena have taken place. Also, the 'Boggart Stones' mentioned in the text (Boggart, remember, being a local term for ghost, spook or spirit) is situated on a fault line.

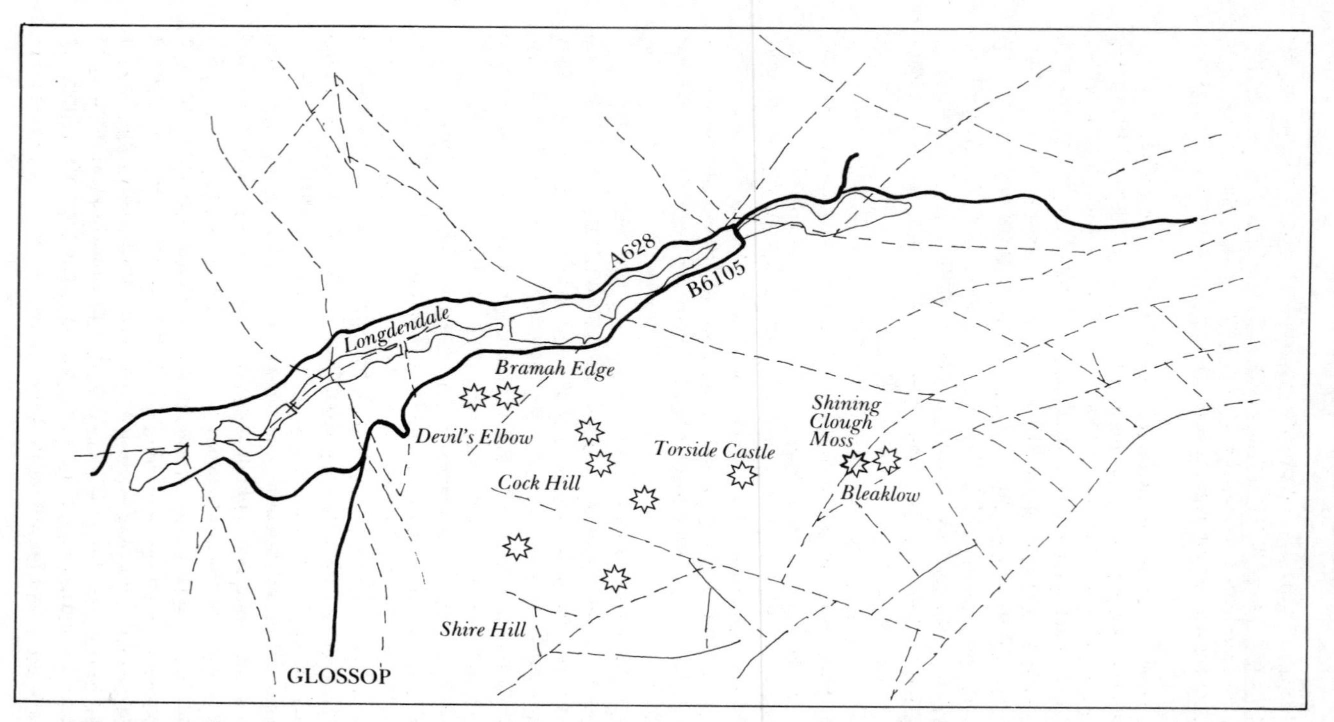

FIGURE 16 Faulting in and around Longdendale, near Glossop, in the Derbyshire High Peak area of the Pennines. Stars indicate some of the locations of repeated reported light phenomena, and faults are represented by broken lines. Note that the string of reservoirs in the valley actually overlay faulting. Shining Clough has been seen to light up completely (see Chapter 4). Distance between Cock Hill and Bleaklow is approximately two and a half miles.

The Project's 'Area 2' contains Buxton in Derbyshire, which was the focus for the early appearances of the 'phantom helicopter' in 1973–4, just prior, the researchers note, to the 24 January, 1974 'meteor' or 'earthquake' geophysical event centred on the Berwyn mountains of North Wales.

Todmorden is one of the towns referred to in the study where various light phenomena have been reported (and also inexplicable sonic effects). The largest fault in the Pennines, the Craven Fault, runs underneath this location. The April 1981 UFO flap around Todmorden culminated on 20 April, coincident with a small slippage in the Craven Fault.

There are specific examples of lights on the Pennines which can be clearly related to seismic activity. During the winter of 1750, for instance, balls of white light making a 'roaring noise' were seen over the whole Pennine chain, within minutes of earth tremors being felt. Five years later, tremors were followed by 'uncommon phenomena in the air: a large luminous body, bent like a crescent, which stretched itself over the heavens'. Reports of other luminous aerial phenomena were made from Keighley, Steeton, Silsden and Skipton at the time. Tremors in these areas of the Craven faults also coincided with the appearance of 'clouds of solid, black material' which wandered above the moorlands in a seemingly controlled manner. These things sound rather similar to the totally black shape which crossed the road in front of John Davies at the Devil's Elbow (see page 90).

Modern times also provide evidence for an apparently clear temporal association between seismicity and phenomena on the Pennines. Between 21 and 24 July 1984 literally hundreds of people across West Yorkshire saw unexplained lights moving about at low altitude. Only a few examples are given in the text in Chapter 4. The mini-wave peaked on 23 July. Local researcher, and editor of *Earth* magazine, Paul Bennett noted that 'of the accounts I have looked at relating to July 23, most of the manifestations observed conform remarkably in their flight paths with the earth faults immediately below them.' One example he gives is the observation of the light observed by Ian Tilleard and his girlfriend over Wrose, near Shipley. Figure 17 shows how the course of the light mimicked the layout of faulting in the area. Another of Bennett's examples is shown in Figure 18, where the course of the Calverley Woods light can be seen following faulting. This wave, of course, immediately followed the large Porthmadog 5.5 Richter quake of 19 July 1984, mentioned above – an exceptional UFO wave after an exceptional seismic event. It may be thought that an event centred in North Wales is too far away to affect faulting in Yorkshire, but a glance at Figure 19 shows that the perceived influence of the quake (from which there were substantial aftershocks – 45 low-magnitude tremors were registered between 19 July and 26 November) stretched well beyond the Pennines. As we will see below, in the work of Persinger and Derr, there is strong statis-

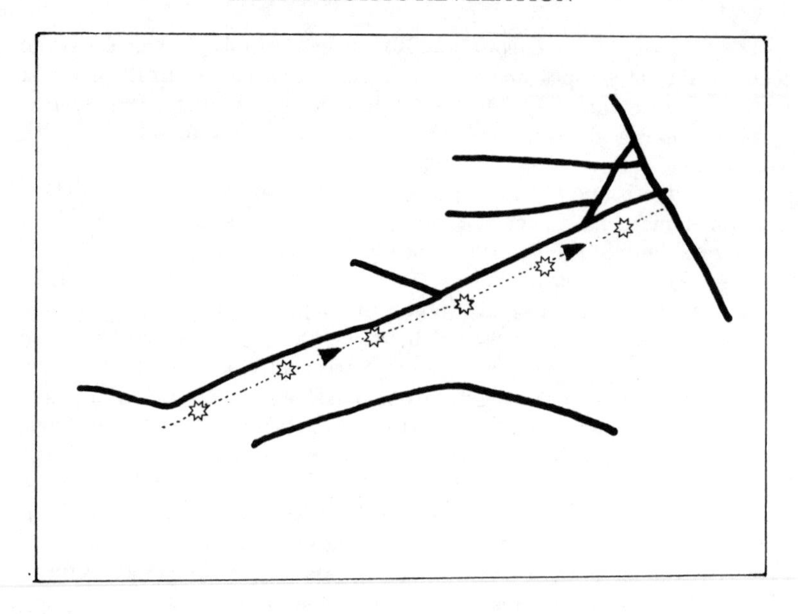

FIGURE 17 Researcher Paul Bennett discovered that a lightball seen travelling from Wrose to Thackley, near Shipley, during the July 1984 UFO outbreak, followed a course adjacent to local faulting. The light disappeared where one fault crossed the course of the other it was following. In diagram, faulting is shown by bold lines, and the course of the light by white star symbols and dotted line. (*After P. Bennett.*)

tical evidence to support the observation that seismic events at some distance from a region can still provide a trigger for the appearance of lights in that region.

In the Area 5 section of Chapter 4, Carleton Moor, southwest of Skipton, was singled out as it had produced so many reports of low-level luminous bodies, a number of which were witnessed by local police and photographed. The moorland area consists of a high anticlinal land-ridge constituted by a layer of gritstone rock beneath which is limestone and shales extending north from the Rossendale valley in Lancashire. The Carleton region is also immediately on the border between the occurrence of gritstone and limestone rock, which occurs to the north of the South Craven Fault, near Skipton. The area is heavily faulted, with fault lines encircling the edges of the gritstone on Carleton Moor and Lothersdale, where there are many disused quarries.

The BGS records for 1967–78 and 1982–4 show at least 25 epicentres of earthquakes with magnitude 1.0 or greater in the Pennine region covered in Chapter 4, showing the area to be subject to ongoing tectonic stress.

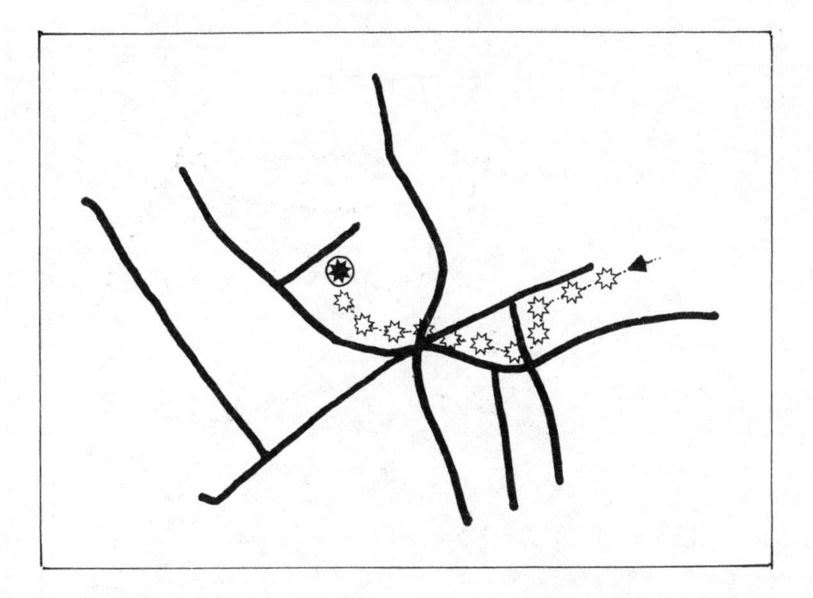

FIGURE 18 Paul Bennett found that the course of the July 1984 Calverley Wood UFO also followed the line of local faulting. Again, diagram shows faulting as bold lines, and course of UFO with white star symbols and dotted line. The light 'landed' at the circled black star symbol position. (*After P. Bennett and S. Hart.*)

THE UNITED STATES

In Chapter 5 we looked at a range of the so-called 'spooklights' in America. They were not specially selected – I included accounts of all those for which I had some information. Figure 20 shows most of the spooklights locations in relation to centres of mid-range earthquake events in the USA. The Illiamna lights of Alaska are, of course, outside the range of the map. I have omitted two or three locations mentioned in Chapter 5 as I was unable to determine their location with sufficient accuracy. The remaining 16 spooklight locations are indicated on the map. It is effectively a random sample. We find that nine of the locations (56 per cent of the sample) occur precisely at, or within a few miles of where quake epicentres have been recorded at some time. Twelve locations (75 per cent of the sample) are situated at, or within 50 miles (80 km) of such points – well within the sphere of seismic influence, as we have seen even in the miniature landscape of Britain.

Although off this map, we might note here that the Alaskan Iliamna mountains – near Lake Iliamna – which sometimes display light

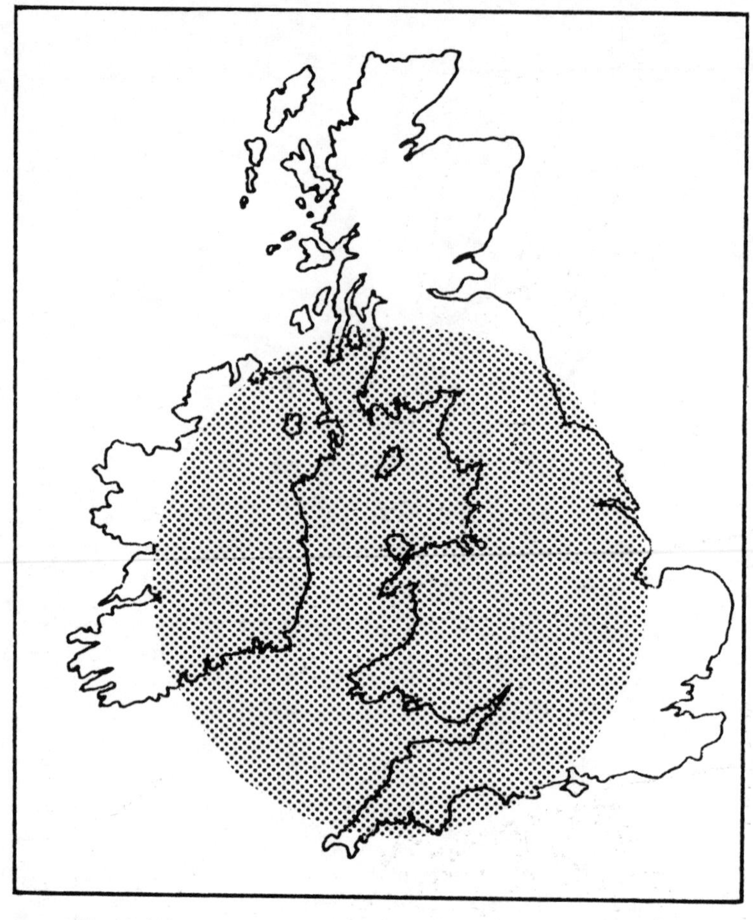

FIGURE 19 Toned circle represents area of perceived effects of the 19 July 1984 earthquake, 5.5 on the Richter scale, centred on the Lleyn Peninsula, Wales.

phenomena, are only about 200 miles (322 km) west of the centre, in Prince William Sound, of the absolutely colossal earthquake (8.5 on the Richter scale) which occurred in 1964. The city of Kodiak, on Kodiak Island nearly 300 miles (483 km) to the southwest, was destroyed by a giant flood wave. As luck would have it, an instrument measuring the magnetic field for an oil company had been left working in safety on a hill throughout this awesome geophysical event. Helmut Tributsch described what it recorded:

> Afterwards, the analysis of the recording trace showed that the magnetic field had increased considerably in strength before the earthquake in several brief,

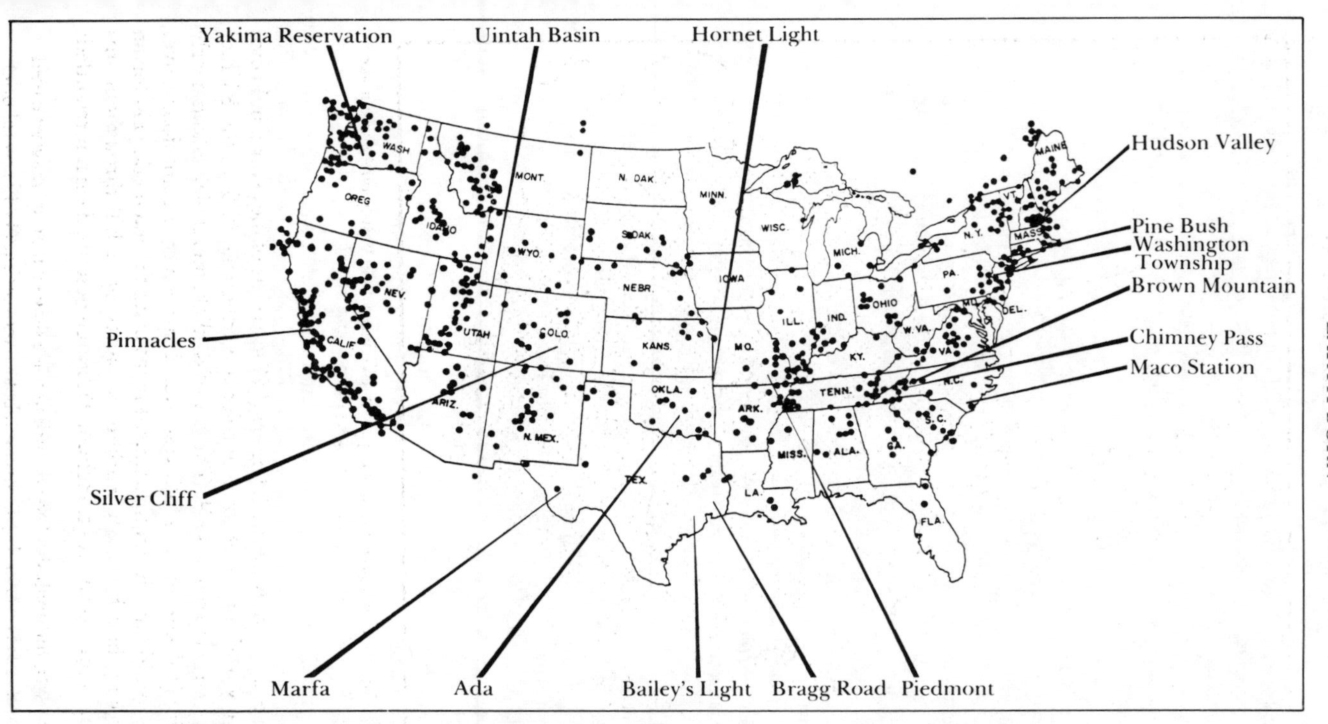

FIGURE 20 Locations of 16 light phenomena locations described in Chapter 5 positioned on map of the USA showing medium intensity quake epicentres (black dots).

closely spaced spurts. The strongest of these pulse-like magnetic-field deviations happened 66 minutes before the onset of the earthquake and reached the astonishing value of 100 gamma. No one knows yet what the mechanism is that led to these short increases in the magnetic field. It is tempting to think of a piezomagnetic effect – the generation of a magnetic field through changes of pressure in the rock.[7]

It is probably not without significance that an area where such fierce seismicity can occur was the centre of a very well-attested incident of light phenomena. On 22 January 1950 two orange balls of light, moving around a common centre, were seen on the radar of a US Navy patrol plane, apparently scrambled the radar at Kodiak, and were observed visually by the plane's pilot and crew and officers on the USS *Tillamock*. The official FBI report of the case concluded that 'the objects must be regarded as phenomena . . . the exact nature of which could not be determined'.

The southern reaches of Alaska – where Illiamna, Kodiak and Prince William Sound are all located – are on the Pacific Ocean's volcanic 'ring of fire', denoting the presence of a tectonic plate margin.

It is reported[8] that the over 100 spooklight locations in the USA have been checked with regard to faulting by the Vestigia group (Chapter 5). They found that all but one 'lay over or near geological fault zones'. If this is the case, it is only natural that Figure 20 should show such a pronounced correlative trend between locations of lights and quake centres.

There are odd scraps of additional information about some of the spooklight areas that further support geological associations. Marfa is adjacent to a recorded centre of seismicity in a huge area where no other events are noted. We can also recall that the best account of the lights there has been given by two *geologists*. They were 'surveying uranium deposits around the Chinati Mountains' and one of the geologists, Elwood Wright, commented that, 'The country's mineralised.'[9] The strange light on the track near Maco Station, Wilmington, North Carolina, demonstrated a pronounced response to seismicity: in 1873 the single light there was joined by a second one (Chapter 5), but a short time after an earthquake in 1886 both lights disappeared completely. After a while, a single light returned to its nightly jaunts. The curious happenings reported last century around Chimney Rock Pass, also in North Carolina, occurred near Shaking Bald Mountain, apparently so named because of the subterranean rumbling sounds that can sometimes be heard there, and the 'almost continuous' earthquake activity that occurs for periods in the area.[10] The third recorded earth lights location in North Carolina is the Brown Mountain area; the Grandfather Fault runs beneath Brown Mountain. The UFO flap around Piedmont, Missouri, was described in Chapter 5, along with mention of the study made of the wave by Dr Harley Rutledge. Apart from the Clearwater Reservoir to the northwest, significant local faulting runs across

Clark National Forest only 9 miles (14 km) southeast of Piedmont, and close to Ellsinore. (As recorded in Chapter 5, Rutledge saw his first unambiguously anomalous light of the flap around Clark Mountain, while in a light aircraft.) It seems the main light phenomena associated with this outbreak are likely to prove of tectonic origin in some way, according to Michael Persinger. He finished analysing the Rutledge data base early in 1988 and noted 'the temporal and spatial distributions of his 1973 observations look very much like a type of earthquake light.'[11]

The Pinnacles light, California, was included as there are two traceable eyewitnesses, one of whom, David Kubrin, has a scientific background, and a previously unpublished [12] photograph of the phenomenon exists. The position of the observed and photographed light must have been within a mile and a half (2.4 km) of the Pinnacles Fault, and no more than 6 miles (9.6 km) from the mighty San Andreas Fault. This latter is one of the most significant in the world, running for some 650 miles (1046 km) through California cutting a suture up to 30 miles (48 km) deep. The Pinnacles are located in one of the most active lengths of the San Andreas.

Two zones of light phenomena have received detailed study by John Derr and Michael Persinger – the Uintah Basin, Utah,[13] and the Yakima Reservation area, Washington State.[14]

Persinger and Derr took the Uintah Basin data base from Frank Salisbury's *The Utah UFO Display*, studying all the reports between 1965 and 1971. The strongest year for sightings was 1967; 36 reports were recorded in the data base for that year. Fifty-four per cent of these sightings were of orange-red light balls and 31 per cent were of other kinds of lights. The rest were of metallic-looking disc shapes and one 'close encounter'. So 1967 was taken as the optimum year for analysis. Statistical techniques were used to study the sightings against 26 earthquakes recorded up to 155 miles (250 km) from the Uintah Basin during 1967. One quake occurred near Roosevelt, the region of the Basin which produced most sighting reports, and the others were scattered around the general area of the Basin. The reports were handled as both total numbers per month and total numbers of days per month in which a report had been made while the earthquake data were assembled as sums of quake magnitudes per month.

It was found that total numbers of reported sightings peaked in February, April and, to their greatest extent, in October 1967. These very closely matched the peak total magnitudes of earthquakes within 93 miles (150 km) radius of the Roosevelt area. Quakes between 93 and 155 miles (150–250 km) did not correlate significantly. Numbers of UFO reports were highly correlated with the magnitude of seismic activity in the same month.

Persinger pictures a *strain field* moving through the crust between shifting centres of earthquake activity. Although this is hypothesised – such a field has not yet been measured – it is really only to be expected that strain must be passing though the Earth's surface during the flexing of the crust in a particular area during times of seismic unrest. Persinger believes that a field accompanying this moving strain provides some kind of geomagnetic factor that triggers the production of light phenomena. It is certainly the case in the Uintah Basin in 1967 that most sightings occurred when such a strain field was likely to be passing through the area where the reports were generated, on its way, as it were, from one epicentre to another. The sightings happened between epicentres in terms of both time and geography.

The mechanism envisaged is, essentially, that the passage of strain triggers light phenomena in strong, resistant geology in a region, and quakes in weaker parts of the area.

Derr and Persinger applied similar techniques to the Yakima Indian Reservation in southern Washington (Chapter 5). They assembled a sightings data base that amounted to 82 reports between July and April 1977. These included a set of observations made by scientists and engineers who came to the fire lookout towers specifically to observe and photograph light phenomena. Seventy-eight per cent of the reports involved nocturnal lights. It was noted that most of the lights were seen in the vicinity of the ridges in the area, with a distinct cluster around Satus Peak, the general area of a surface rupture and one of the stronger earthquakes in the area during 13 years covered by the study. The Toppenish and Ahtanum ridges are still undergoing compressional deformation – a 20-mile (32-km) length of the Toppenish Ridge fold has nearly 100 surface faults and experiences shallow quakes. Nevertheless, seismicity for the period of the sightings was low. It was felt that if the pattern of seismicity was similar to that in a nearby but much better studied area, then there would be swarms of very small seismic events as well as the more distinct quakes. Unfortunately, there was a complete absence of seismograph stations in the area where lights were reported, so this possibility could not be tested.

The first general check between quake energy and sighting showed that seismic events over 62 miles (100 km) away did not correlate, but closer activity did show increases in the two periods of most sightings in 1972 and 1976, and there was a long period of successive reporting of lights for seven months prior to the greatest release of seismic energy in the studied period (June–July 1975). There was not consistent matching of sightings with seismicity for the entire 1972–7 period, but studies of certain aspects of the data over specific time periods did reveal some closely similar patterns. One problem felt by Derr and Persinger was that much of the seismic data came from events occurring at long distances from the Satus Peak centre of light activity on the reservation. So they looked for a period when epicentres occurred within 50 km of the peak. One month, October–November 1976, met this criterion, and it precisely matched the period with the largest numbers of light phenomena that had ever been reported in the area.

The investigators also searched for evidence of a moving stress field. The direction of tectonic stress in the Toppenish area is north–south compression with an active fault running east–west through the middle of the area generating sightings of lights. From June 1976 to March 1977, Derr and Persinger noted 11 earthquake-light phenomena cycles 8 of which occurred in time intervals between quakes located north and south of the sightings and 2 more occurred on the days when there were north–south shifts in epicentres. This strongly supports the idea of a moving 'band' of stress within the local geology with an associated field effect triggering light phenomena. As this theory requires a change in geomagnetism, Derr and Persinger looked for days when the magnetic index was exceptionally high. They found some evidence to suggest that such maxima *precede* the occurrence of light phenomena, but the picture is by no means clear as yet.

In the conclusion to their Yakima study, the two researchers commented:

> We are not suggesting that geomagnetic activity might be a causative factor in triggering earthquakes. Rather, we consider it possible that earthquakes and geomagnetic activity are independent processes which, when they happen to be in a certain phase relationship with each other, may locally enhance conditions for producing light phenomena . . . .
>
> Light phenomena tended to occur when there was shift in the location of the closest earthquake activity, particularly if they moved across the observation area, which essentially coincides with the Yakima Indian reservation . . . the spatial-temporal pattern may reflect some process associated with changing tectonic stresses across the reservation. Light phenomena would be natural phenomena coupled to these local stresses, which are almost invariably compressional wherever these apparent geophysical luminosities have been observed in the U.S. This hypothesis is also consistent with the possibility that light phenomena are caused directly by very small quakes below the threshold for location.

An extended data base used by Derr and Persinger also suggested that there was a build-up in the number of sightings prior to the eruption of Mount St Helens, and a decrease since that event, though this apparent trend has yet to be confirmed.

WORLDWIDE

With only patchy research in fairly well studied areas like Britain and America, it is to be expected that earth lights work on the widely scattered examples referred to in Chapter 6 will be minimal. The exception, however, is Scandinavia; Dan Mattsson has done sterling work in Sweden both single handedly and in conjunction with Michael Persinger, and, of course, there is Hessdalen. Old mines in and around Hessdalen give a clue, and Hilary Evans states it bluntly:

> Hessdalen is a geologist's dream: the land is stuffed full of minerals of many kinds, and copper mining was once carried out nearby. The magnetic field is the strongest in the whole of Norway.[15]

As Mattsson has pointed out to me in personal communication: 'Hessdalen is situated in the middle of one of the largest ore-fields in Norway (copper and zinc). The UFOs appear and disappear where the magnetic fields are strong.'

It seems there was little available geological information on Hessdalen, so the Project team obtained aerial stereo photographs which were submitted for professional analysis. Faulting in the hills around Hessdalen was confirmed.

Evaluating seismicity has been a problem: Hessdalen is not well covered by seismographic instrumentation, as the Project's *Technical Report* indicates:

> There has to be a quake with a greater magnitude than 2.5 on [the] Richter scale in the Hessdalen area, to be detected on one of the nearest stations.
>
> Inside a radius of 70 km from Hessdalen, there have been detected four quakes, all small in amplitude, during the last six years. Inside a radius of 50 km, there has been 15 recordings, during the last 100 years. Even though the area hasn't got any good cover, you can say that there is very little seismological activity, compared with other areas in Norway.[16]

This last opinion seems to be based on the assumption that seismicity for an earth lights area would need to be a highly localised quake of substantial magnitude. As we have seen, this need not be the case. Sources of seismic disturbance up to certain distances can still affect geological factors in a locality, and, as in the case of Hessdalen particularly, very low-magnitude seismicity locally can pass undetected. It is also of possible significance that of the 14 seismic events within 50 km of Hessdalen listed in the technical report in the last hundred years (only 14 listed, even though 15 are claimed above), three of them (21 per cent) occurred between 1980 and 1982. Not only had three quakes not been previously recorded as occurring so closely together in time, the lights started their appearances in 1981, in the middle of the activity. And, as we will see shortly in Sweden, areas that resist quake activity themselves for long periods can succumb to tectonic disturbance eventually: it is almost certain that such areas experience strong strains within their stubborn geology prior to energy release in an earthquake.

The project installed a seismograph at Hessdalen on 24 October, 1983, to cover the winter session. It seems that regular study of the instrument only took place in February 1984, during the intensive winter field study at

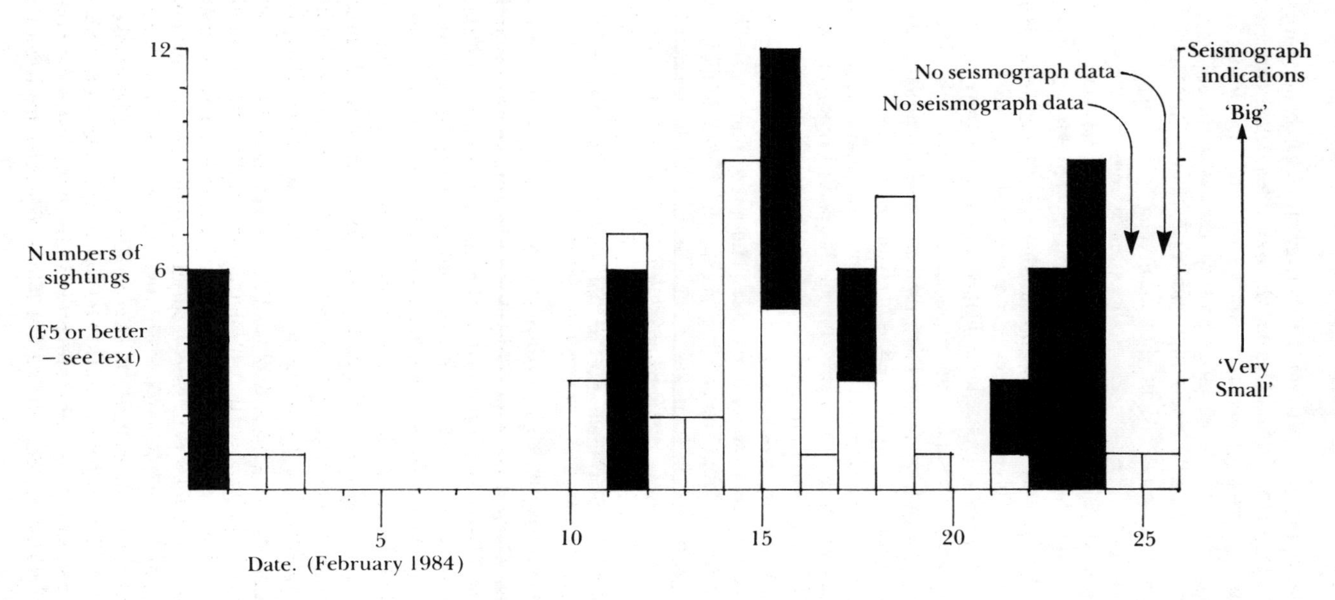

FIGURE 21 Histogram showing the relationship between observed high strangeness light phenomena and recorded seismograph indications at Hessdalen, Norway, between 1–16 February 1984. Sightings of lights are represented by white columns, and seismograph indications by black columns. Columns are mixed black-white appropriately for days on which both lights and seismic events occurred. It can be seen that the biggest observed seismograph indication occurred approximately in the middle of a wave of increased light phenomena sightings around mid-February.

Hessdalen. Twelve readings between 1 February and 24 February are the only ones presented in the *Technical Report*, when the Hessdalen team 'could clearly read visible recordings at these times'. Quantified readings seem not have been made, and the seismograph responses were simply referred to as 'very small', 'small', 'medium' and 'big'. These are, however, sufficient to give us some insight into local seismicity when studying sightings for the period. Figure 21 provides an extremely rare piece of information: immediately local seismographic indications taken at the time observations of light phenomena were being made. The Hessdalen team employed a graded system regarding the sighting – the more unusual, inexplicable lights were given a higher 'strangeness rating', which, in their notation, meant that F1 could be identified as a mundane source and F10 was almost certainly an unknown phenomenon. One of the Project directors, Erling Strand, recommended that only sightings of F5 strangeness or higher should be considered as likely to be genuine unexplained lights. Following that advice, I used only F5 or higher sightings in producing Figure 21. Now, it was stated in the Hessdalen report that there was no connection between the lights and the seismograph results, a statement that has been repeated by others subsequently. A glance at Figure 21 shows that this view is not tenable. There is a *distinct* correlation between the two sets of data. The one 'big' seismograph reading took place on 16 February, and it can be seen how the number of F5 or better sightings dramatically increase from 12 February, peaking on 15 February and falling away by 20 February. If light phenomena do respond to local strain, then Figure 21 is a perfect illustration of that – a classic expression of the earth light thesis.

It seems that because the origins of the local disturbances recorded by the Hessdalen instrument were supposedly distance, the information was disregarded and never studied in visual form. The point is, some level of disturbance *did* occur at Hessdalen in conjunction with an increase of the better sightings. This seems to have been subsequently realised to some extent, as Erling Strand, writing in April 1986, stated:

> all movements recorded had their centres in other parts of the world. But we did not measure strain. Strain usually occurs before the movement (or quake). Now we have recorded some small quakes in the Hessdalen area: maybe there were strains in the area at the time the UFOs were observed?[17]

Writing in the specialist UFO journal *Magonia* (April 1984), Jan Krogh of NIVFO (Chapter 6) expressed the conviction of himself and his organisation that the lights at Hessdalen were a natural phenomenon. During their careful questioning of witnesses to some of the more interesting reported events, they found that revealing phrases were used. Lars Lillevold, for example, who some writers claimed had seen a 'structured' object near his farm, actually saw something that resembled 'burning gas'

– it did not even present a metallic appearance, much less 'windows' or other similar features.

Before leaving Hessdalen, there is one extraordinary, and possibly vital observation made at Hessdalen to note: 'project members sometimes reported a curious "rippling" motion – like on a boat at sea – as the ground appeared to respond to the passage of a UFO.'[18]

In Chapter 6 there are descriptions of the seven Swedish UFO 'windows' identified by Dan Mattsson. We revisit them here, this time filling in their geological and seismic aspects.

Window 1, it will be recalled, was the Gislaved and Gnosjö district, where sightings were well over three times the expected number. This statistic was caused by a steady, low-level occurrence over the years rather than any dramatic outbreak. There was no earthquake in the region until 6 September, 1977 when a strong disturbance occurred (3.0 on the Richter scale). A sighting was made on the evening of the quake, but, Mattsson observes, 'after the quake there have been very few sightings from this district'.[19]

The Kolmarden area near Nyköping and Oxelösund was designated as Window Two, one of the most important UFO districts in Sweden, where four times the average number of sightings has been recorded – a consistently high rate for 20 years. The area is situated along the Braviken Fault, one of Sweden's main fault lines. It is also a highly mineralised area. Compass deviations occur in the Baltic off Oxelösund caused by a large body of ore below the sea, which is in the very area where fishermen saw a formation of 16–18 disc shapes which changed colour. Mattsson's studies of the area showed that out of 44 sightings, 39 were 'closely aligned to magnetic ores' or positioned within a zone of magnetic disturbance existing in the eastern part of the area. One light actually appeared over a disused mine. Iron, copper and cobalt were mined in the district – Oxelösund has iron works – and there are also deposits of lead, manganese and sulphur (as at Hessdalen). Mattsson reports that 'gamma radiation from the bed-rock is intense'. As I have remarked earlier, I suspect that high background radiation is a possible contributory element in the production of light phenomena in certain circumstances.

The Köping, Arboga and Kungsör region, comprising Window Three, displayed eight times the average number of sightings for a period, but ceased to do so after 1978. The nearest earthquake to be registered to this location was 25 miles (40 km) away.

Mattsson stated that the Vallentuna area (Window Four) provided about four times the average rate of sightings, most deriving from 'the most intense flap yet in Sweden' in 1973–4. Some of these sightings, including curious beams of light, were described in Chapter 6. There had never been any recorded earthquake in this area until a surprising event on 23 December, 1979, when one of the most powerful Swedish quakes ever known took place there. This led Mattsson to comment: 'This indicates that there can be very powerful stresses built up in areas where there has never been any earthquake registered. . . . Is the bedrock and faults more stable here and the tectonic stress higher?'

The fifth window, Dalarna, has double the average number of sightings. Mattsson notes that there have been no recorded quakes here, but that 'the sightings follow, exactly, the mining district of Dalarna with a wealth of different minerals.'

Window Six, the Sundsvall and Härnösand district, registers two and a half times the expected level of sightings. Sundsvall is an old earthquake centre, and it has the remains of an old volcano. The region has an active UFO group, and from these people Mattsson was able to obtain 54 sightings from a 20-year period. These sightings had been filtered by the local researchers, with obvious misidentifications removed. Mattsson then compared these to data on 14 quakes in the area. The correlation is very strong, as can be seen from Figure 22, compiled from Mattsson's findings.

The seventh window, around Kiruna and Gällivare, is the main mining district of Sweden, with many earthquakes.

In 1986, Persinger worked from Mattsson's data base and analysed sighting reports in Sweden between 1963 and 1978. These were then compared with the 147 (relatively small) earthquakes recorded for the period. The results of the statistical analysis were similar to findings made by Persinger in previous studies: there was an increase in sightings prior to an increase in earthquake activity in the *following* six-month increment. The rise in sighting preceded a rise in seismicity.[20]

While the available Scandinavian data strongly support the earth lights thesis in several different ways, there is little relevant information about sites elsewhere in the world, so only a few general comments can be made here regarding some of the locations mentioned in Chapter 6.

The lights in the north China Sea and in and around Japan are not difficult to associate with geological factors, for instance, as they occur close to tectonic plate margins and the consequent 'ring of fire'.

No one to my knowledge has made a serious study of possible geological and seismic associations with min min lights in Australia, but in 1983 I did ask one of my correspondents, George Sandwith, to have a preliminary look at this for me. Sandwith is fairly well acquainted with the Simpson Desert area of western Queensland, and he studied geological maps of the region, bearing in mind reported min min light events he knew of. He found that there was a fault line near most of the reported locations, stating: 'To my mind, the proximity of the fault is most convincing.' With regard to the light phenomena reported by the Knowles family and others on the Nullarbor Plain in southern Australia near the Great Australian Bight, *Nexus* magazine made the observation that 'An interesting coincidence is that 24 hours later the greatest series of earthquakes in Australia's history

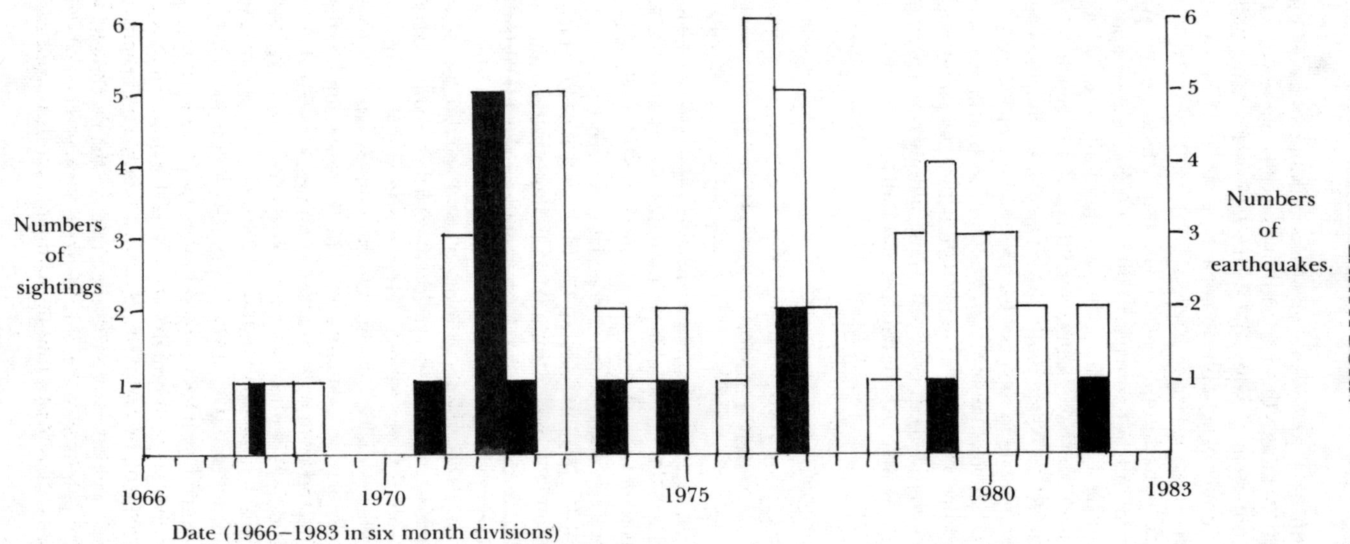

FIGURE 22 Swedish researcher Dan Mattsson studied the relationship between reported light phenomena and recorded seismic activity at various Swedish UFO 'windows'. This histogram is based on his data for a number of years up to 1983 in the Sundsvall-Härnösand area. (Black columns represent recorded earthquakes; white columns represent UFO sightings.) His data show that three years preceding 1966 had no recorded UFO or seismic events. Here we see seismic activity 'embedded' in all periods of reported UFO activity, interspersed with periods having neither UFOs nor quakes.

began near Tennant Creek.'[21] When Alanna Moore passed through Tennant Creek (in Northern Queensland) in May 1988 she noted (*Dowsing News*, 17, 1988) that the local paper, the *Barkley Regional* (25 May, 1985) was reporting the occurance of light phenomena there, around 'the town where earthquakes were not supposed to happen'.

Hawaii provides a backdrop of active volcanic activity to the Waimea lights; while the lights on the pampas containing the Nasca lines, and indeed any of the Andes light phenomena, are automatically granted tectonic associations, as the Andes have been thrown up on the margin of a tectonic plate. There is, in fact, a 'Nasca Plate', which is busy forcing itself beneath the South American Plate to the east. On the east side of South America, the only geological factor known for sure about the Brazilian location of Cynthia Newby Luce's 'mother of gold' light is that the property is on a mountain comprised mainly of granite. Minor faulting could probably be inferred from this, as well as high background radiation.

HOLY MOUNTAINS

The same situation applies to the mountains and hills around which lights have been reported – little is known. Mountains are usually markers of some form of tectonic disturbance, however, as well as possessing the antennae-like nature of their peaks. Cader Idris, where I personally witnessed occurrence of a rapidly moving light in 1982, is situated over the major Bala Fault, and the rocks forming the mountain are volcanic. I have no direct information on Changkat Asah, home for the Malayian lights described by Sir George Maxwell, other than the observations that the area lies on the ring of active volcanoes around the Pacific, thus close to a tectonic plate margin, and that tin was dredged from swamplands in the Malaysian peninsula. Mount Athos in Greece is in one of Europe's most intense regions of seismicity. Mount Omberg in Sweden is probably granitic, with possible local faulting and high background radiation. Pendle Hill in England is an outlier of the Pennines, composed chiefly of gritstone, underlaid by limestone and shale. The hill contains a number of faults having a considerable downthrow to the southwest through Barley Moor and the Nick of Pendle, and to the north near Clitheroe. It was noted in Chapter 6 how a sighting in 1869 was coincident with an earthquake. No faulting is recorded at or immediately around Glastonbury Tor in Somerset, but the line of a recorded fault to the northeast can be projected through the hill to meet up with a recorded fault to the southwest, which follows the same approximate axis. It is likely that geologically recent material overlays the older rocks containing the fault (Glastonbury was an island in the shallow waters that occupied the Somerset Levels in late prehistory). But in any event, we know that the hill has been prone to seismic events, as a church on the hill's summit was destroyed by an earthquake in the medieval period. And, half the world away, California's Mount Shasta is, of course, part of the same mountain range and on the same tectonic plate margin as were Arnold's flying discs.

LIGHTS IN THE LABORATORY

I would have though that enough evidence has been presented so far in this book to convince anyone that some kind of geological association with light phenomena is, at the very least, probable. It is beyond the space available in this present work to display all the evidence for the the theory: Persinger, particularly, has produced a formidable number of papers on aspects of the tectonic strain hypothesis, studying numerous data bases, both historical and contemporary, in North America and Europe. Resistance to the theory on the part of critics is primarily based on either ignorance (in the sense of having absorbed only part of the evidence available) or personal preferences and attitudes.

The link between landscape and lights is a major first step in the understanding of the whole enigma. But while it has proved possible in this book to present a reasonably detailed picture of the matrix of forces and factors that seem necessary for earth lights to appear, the actual mechanism whereby the lightforms manifest is not understood as yet, and neither is the precise nature of the energy comprising the lights. Some critics consider that the theory cannot have any value until these matters are known. But the mechanics and nature of the lights are the *goals* of the research, not something that happens on the way. To dismiss research that is leading us towards those goals is completely illogical, like dismissing the role an egg has in the development of a chicken. To dismiss the presence of a phenomenon just because we do not know its full nature is like saying that lightning did not exist until we understood the workings of electricity. It is perfectly respectable to conduct research into something without fully understanding its nature - science does that in many areas all the time! If this was not the case, there would be little need for research or scope for the advancement of knowledge.

But if stress and pressure on rocks is a key factor in the production of earth lights, it might reasonably be assumed that the process can be approximated in the laboratory. It seems this might well be the case, though we should always be aware of the limitations presented by the relative simplicity and tiny scale of any laboratory set-up compared with the vast and complex canvas of nature – only conditions known and understood can be provided as the initial framework in laboratory models, while there are clearly unknown aspects to the actual, landscape manifestation of earth lights. Nevertheless, the 1981 observation by Dr Brian Brady – of the US Bureau of Mines in Colorado – of lights associated with the fracturing of rock during tests in rock crushers, provided a possible bridge between landscape and laboratory. Using slow-motion photography, Brady was able to show that small lights with brief lives of up to a second could be observed flitting around a rock crusher's chamber moments before rock fractured. The lights do indeed look like mini-UFOs (Plate 32).

In London in 1983 Paul McCartney and I, through the good offices of colleague John Merron, were able to conduct privately a similar experiment to Brady's. We used Cornish granite for our experiment. This was in the form of cores which had been drilled for other purposes. These were cut to various lengths up to 6 inches (15 cm). Their ends were polished smooth and the cores were placed, one at a time, between the massive, horizontal metal plates of a rock crusher. This machine could gradually exert increasing pressure on the column of granite, until it gave way explosively, or it could apply sudden pressure to a core. We used both modes in several runs with the cores. The eyes of observers were protected (theoretically, at least) from flying rock fragments by clear plastic screens drawn across the chamber containing the core.

We had an initial run to check things out before blackout conditions were applied. Even in broad daylight we could see a brilliant orange light, appearing at least a centimetre across, flash out from the core. Subsequent runs took place in blackout conditions. The lights looked magical, occurring less than 2 feet (60 cm) from us. I saw three distinct types of light. The most common were balls of orange light. These apparently darted out from the core, dashing around the chamber. In a few instances, diffuse curtains of pale orange luminosity would appear. But on just one occasion, I saw a beautiful, vivid blue light. It was smaller than the others, and was longer lived, lasting for at least one second. It twinkled as it described a seemingly controlled horizontal arc through space an inch or two from the surface of the granite core – the distance kept as regular as if the light was on a wire extending from the rock. Then the core broke, and the light disappeared.

Plates 33–7 show aspects of the rock-crush experiment, and Plate 33 is an open-shutter picture obtained by John Merron of the path of a light in the rock crusher chamber. This might not be the picture of an actual rock fracture light, though, because Brian Brady and Glen Rowell obtained a similar picture in a 1986 experiment (see below). While they confirmed that authentic rock fractures lights can be observed in darkened conditions, they were able to show that their photographed lights were burning chips of steel scoured out of the rock crusher's plates by exploding rock fragments. So at least some of the brilliant orange lights I have described may have been cause by such an effect, but the orange curtains of light I saw, and the intense blue light, were definitely related in some way with the pressured granite cores.

Superficially at least, the only apparent difference between these laboratory lights and earth lights or UFOs seems to be one of longevity and scale. This is only to be expected – a rock core a few inches long hardly corresponds to miles of complex geology in the landscape.

Up to this time some kind of mechanism based on piezo-electricity (Chapter 1) was broadly assumed for the production of the lights. Persinger had put it forward in 1977 as one *possible* mechanism. And, indeed, it is easy to see how this effect may well be one of the factors involved in the complex of conditions that allow the appearance of earth lights. But it seemed increasingly unsatisfactory as the main 'motor' behind earth light appearances. McCartney and I looked at a similar mechanism of releasing luminescent energy from rock – triboluminescence (the production of light by means of friction). It is a well known but incompletely understood phenomenon that can be demonstrated as a party trick: if a person cracks a sugar cube between his or her teeth in a dark room, the ensuing flash of light from the mouth will usually cause astonishment in an audience. Or two pieces of quartz rubbed together can light up a darkened room like candlelight. Some Amerindian shamans used to produce a 'magic fire' by shaking quartz pieces in a leather rattle device containing holes – the friction caused by the shaking created light from the crystal pieces which shone out through the holes.

McCartney and I carried out some *ad hoc* experiments, and pictures of some of the triboluminescence effects we produced can be seen in Plates 34–7. We found blocks of rose quartz six inches (15 cm) long would glow momentarily after friction with another piece of quartz – it is not a subtle effect. Studies of various friction experiments using standard and infra-red film revealed points of light apparently above as well as on the surface of the rock, glowing amongst the dust produced in the friction. Carrying out the same type of experiments in a closed environment enriched with negative ions seemed to enhance the glowing effects, and we also found that the light can be produced with brilliance under water.

Friction, pressure, heat – these are some of the more obvious ways in which rock can produce light. The basics of what seems to be happening is that electrons are discharged from rocks by such stimuli – this is called *exoelectron* activity. This happens because most minerals are reservoirs of free electrons, trapped at points within the mineral lattice. These unattached electrons are produced in the rock due to the action of natural radiation over geologic periods of time knocking them out of their atomic orbits. They fill the gaps within the mineral's lattice structure until such time as some mechanism discharges them. Friction causes the electrons to 'jump' from higher to lower energy levels and light is emitted in that process. That electrons are emitted from rocks under appropriate conditions is certain: as noted in Chapter 1, Stuart Hoenig of the University of Arizona has measured such emission, giving a basis for Tributsch's idea that animal behaviour is affected by an impending earthquake because of electrical charge in the atmosphere. But it is still not clear, as Hoenig remarks, why rock should give off electrons when it breaks, nor how they actually escape.[22]

We voiced our doubts about piezo-electricity in a *New Scientist* article (1 September 1983) and gave a brief account of the triboluminescence experiments. But as Dr K. V. Ettinger of the University of Aberdeen wrote in response to our article: 'I cannot imagine any process related to piezo-luminescence which will generate lights high in the air.'

In 1984 I gave a talk to the Physical Society at the University of Sussex. Afterwards, one researcher there presented remarkable slides taken with light intensifiers showing feathery light effects emerging from rock under bombardment. The interesting aspect to this work was that light phenomena were produced in vacuum conditions, and generated from *non-piezo* rocks. While piezo-electricity undoubtedly occurs in the landscape under certain circumstances, and in some cases may set loose processes that culminate in light phenomena, it was becoming clear that it could not be directly responsible for the production of lights, or, at least, there had to be something else at the heart of it all. Since about 1980 Persinger, too, has avoided emphasising piezo-electricity, but popular comprehension of his work is still frequently plagued by the association – phrases like 'Persinger's piezo-electric theory' being typical. He realises more than anyone that until we understand how stress in the Earth's crust is actually translated into transient, focalised energy displays ' a variety of models must be investigated . . . . The piezoelectric option is only one possible model.'[23]

In 1986 the results of a fascinating experiment studying rock-crush lights was published in *Nature* by Brian Brady and Glen Rowell. [24] They fractured cores of granite, which has a high quartz content allowing piezo-electric effects, and basalt, containing no piezo-electric minerals, in various atmospheres, in a vacuum, and under water. They used a spectroscope linked with image intensifiers in order to capture the spectra of any fleeting lights produced in the experiment. (Although the light emissions are visible to the naked eye in darkened conditions, occurring as 'discrete pulses' from the rock, light enhancement was necessary so that spectra would show up strongly enough to be photographed.) By studying wavelengths captured by a spectroscope, it is possible to identify individual elements in the energy observed. It is hoped that this experiment would help give some idea as to the nature of the lights.

The results were highly instructive. Both the granite *and* basalt cores produced light. This, of course, argued against a piezo-electrical origin for rock fracture lights, an argument further supported by the lack of any spectra showing ionisation and the absence of microwave emanations. But the biggest surprise was that the spectra (very similar for both the granite and basalt cores) showed no trace of any elements from the rocks, only from the surrounding gas (air, argon and helium were used separately in the experiment) or liquid (water). Spectrum analysis of the lights produced in the vacuum showed only evidence of air (although highly rarified, air molecules will be present to some extent in laboratory vacuum conditions). The spectra showed distinct lines rather than a continuous range of wavelengths (which would be produced if the spectra were due to heat produced by friction in the rock – when they rubbed pieces of quartz together, Brady and Rowell produced a 'smooth' continuous spectrum). Thus rock fracture lights are clearly of a different nature to friction-produced luminosity. And another popular notion, that of the lights being plasmas, was also knocked down, due to this lack of continuous spectra indicating a continuum of radiation, and due to the lack of microwaves. The energy producing the light effects was, in effect, 'invisible'. What could it be? Brady and Rowell concluded that 'an exoelectron excitation of the ambient atmosphere is the mechanism responsible for the light emissions observed during rock fracture.' But this statement in itself, of course, begs a lot of further questions. They openly admitted that, 'An exoelectron excitation type of illumination is not likely to cause isolated light sources to appear several hundred metres above the Earth's surface.'

As John Derr has written:

> Brady and Rowell's work is a significant step to finding one . . . mechanism for the production of geophysical luminosities. But investigation in other areas are still required because we may be looking at several phenomena which sometimes share a common appearance and name.[25]

So while the glimmerings of some kind of mechanism might possibly be coming into view, the full explanation for lightforms like certain kinds of EQLs, quite apart from earth lights or UFOs, is still far from obvious. To understand these relatively stable and remarkable lightforms will necessitate us entering 'a new, challenging area of geophysics' as Derr has so aptly put it. In my view, we are not going to be able to explain them solely within current frameworks of understanding. We may be looking at a potential energy base for the next millennium, our present relationship to it being like that of medieval society with regard to electicity. The current state of human knowledge is undoubtedly going to be extended by the study of earth light phenomena.

BIRTH FROM EARTH

And that study is already throwing up unexpected and exciting possibilities. For example, when Brady and Rowell fractured their rock cores under water, the water glowed and both atomic and molecular hydrogen was produced. The effects on the water were similar to those produced by beta particle irradiation (electron bombardment), and was one of the effects that led the experimenters to propose exoelectron excitation of the ambient medium as the way in which light was produced. But this 'molecular dissociation' has another implication, as the investigators noted, in that it might produce interesting chemical effects. In a discussion of their experiment, they commented that they felt the role of this process 'in initiating chemical reactions of geological and biological interest' was currently being given insufficient consideration. Derr further interpreted this as meaning the process had 'implications for biogenesis', or the origin of life on Earth. Current theories regarding the origin of life usually revolve around notions of lightning activating the primeval muds into producing amino acids, or else extra-terrestrial intervention in the form of bacteria coming to Earth on fragments of meteorites or matter from comets. Now that we know the very rocks of our planet produce energy, we have another potential candidate for initiating the processes that have led to the life cycle. If this is so, then the mysterious lights weaving through our atmosphere have more to tell us about terrestrial than alien life. As the manifestation of earth lights seems to be in some way associated with energy release from the rock body of Earth, then they are, in a sense, ancestor lights.

The creation myths of primary peoples have always maintained that we are Earth born. Mother Earth. Perhaps the scientific study of earth lights is leading us to a renewed appreciation of an archetypal truth. Like glittering Pied Pipers, like planetary messengers, the dancing lights in the sky may be leading us back to the world we so readily disregard – Planet Earth.

8 EARTH LIGHTS AND CONSCIOUSNESS

Geology is thus the gateway into the geophysical arena in which earth lights appear and perform. The geological and seismic connection with terrain-related lights is of crucial importance, for it represents a true starting point for their study. It makes possible the opening of a genuine field of research.

This dawn of realisation regarding the conditions that frame their appearance, however, does not tell us what the lights actually *are*. Here we have to face up to another apparent characteristic of earth lights frequently mentioned in the examples given in Part 2: the seemingly intelligent behaviour of the lightforms. To this, we must add the (admittedly rare) accounts that occur in ufology of abductions, structured craft, physical effects on witnesses, and the observation of strange entities.

But all this cannot be tackled in a single bite – we must progress by stages. First, what possibilities can be entertained as to the nature of the lights? They could be some kind of pure electromagnetic lifeform. C. G. Jung was amongst the first to float this possibility, and there have been others. Indeed, it was the conclusion that Kenneth Arnold came to himself after he had experienced further sightings, and Trevor Constable refers to them as 'critters', claiming to have photographed them on infra-red film![1] There certainly are behavioural aspects of the lights that are reminiscent of the playful quality of a kitten – or even a dolphin.

A more popular idea, of course, is that the lights are alien spacecraft. But the vastness of the universe is such that even allowing for light-speed travel it seems profoundly unlikely that any extra-terrestrial beings have visited our planet by any kind of technology we can envisage. Quite frankly, I do not think that 'nuts and bolts' ETH adherents see how puny their ideas are, nor how conveniently in step with our own cultural development they happen to be – and we saw in Chapter 2 how views of unexplained aerial phenomena have matched the mentality of the day throughout history. However, it is just conceivable that some advanced race in far stellar reaches has such a profound science that it can manipulate natural forces in a way that allows its members to transfer themselves into other places in the universe that would be unreachable by linear travel, however fast that might be. The geological, geophysical conditions which so repeatedly accompany the lights may simply provide a necessary energy window for allowing manifestation by aliens at a location. ETH adherents are quite mistaken not to look at our home planet as possibly containing the basic energy conditions that can be commissioned by ETs in order to facilitate their appearance here. Perhaps the inexplicable behaviour of the lights is the aliens' way of trying to point out that their use of nature for the traversal of vast spatial distances is available to us here on Earth if we had but the eyes to see. That could be why they bring certain types of location to our notice time and again, waiting for us to draw the obvious inferences. The same idea could be applied to inter-dimensional visitation.

If I was to favour one of these notions, it would be the lights-as-creatures possibility. But there is another theory that I continue to find the most persuasive, and which will be pursued here – that earth lights are, in fact, energy forms produced by processes at the very limits of our current scientific comprehension. I think we are looking at an energy manifestation that is either an unfamiliar form of electromagnetism, or else is of a completely unknown order that interacts, resonates, in some way with parts of the electromagnetic spectrum. Such a secret force has long been assumed by traditional societies. In old China it was *ch'i*; to the Australian aborigines it is *kurunba*: a primary sea of force that underpins the manifestation of energy effects and matter in the material world. Either way, the road to its comprehension will have to be through an extended form of geophysics. This energy has remarkable characteristics and has much to teach us.

But how can earth lights as packets of exotic energy explain the structured craft, the humanoids, the abductions?

THE REALM OF THE FABULOUS

We first have to see just what needs to be explained, because the simplistic concept that witnesses see craft with occupants is insufficent. The whole matter is much more confused than that. Many of these reports may not be what they seem. Some are undoubtedly lies or hoaxes. Others are believed to be genuine by the witness, but are likely to be dreams or imaginative episodes so vivid to the person involved that there is difficulty in separating them out from consensus reality: some people are a close encounter just waiting to happen. In *Earth Lights* I gave an Australian case in which a woman genuinely believed she was being periodically taken aboard an ET spacecraft. When accompanied on one occasion by investigators, she experienced being taken aboard an ET spacecraft while everyone else could see her still seated in a parked car! Something had happened to the woman, but it was not a physical visit to a spacecraft. Another case was well researched by Andy Roberts and colleagues in 1988. Alerted to the report of an alien abduction by UFO researcher Jenny Randles, Roberts and friends visited the witness at her home in northern England. The woman was in her forties, clearly distraught by her experience. She felt she had been awakened in her bedroom by two strange beings with 'leathery skin' who 'floated' her to a nearby field where she was taken inside a large spaceship. She underwent the sort of intrusive physical examination widely reported by abductees. She described the craft as being illuminated 'like a fairground', she showed the investigators faint marks on her arm where she said a tube had been inserted, and mentioned a smell like that of cinnamon while events were taking place on the spaceship. She eventually found herself back in her room, and the next day showed physical symptoms she ascribed to the event.

The investigators spoke to the woman only two days after the abduction had supposedly taken place. They were convinced that her emotional distress was genuine. However, Randles felt the case might be suspect as the reported abduction had happened the same night as the British TV showing of an episode of the American soap opera *Dynasty* in which Fallon, a character in the series, was depicted as being abducted by a UFO. The investigators raised this with the woman, who freely admitted she had watched the programme but strongly denied that it had influenced her experience. Nevertheless, Roberts and his colleagues noted that the detailed description of the woman's UFO closely matched that of the TV version, and that the smell of cinnamon and the leathery skin of the aliens were components of the soap opera drama as well. The investigators felt 'the chances of someone *genuinely* being abducted on the same night as the *Dynasty* abduction *and* having an almost identical experience are pretty slim.'[2] They also discovered that the woman is a friend of a UFO 'buff', the person, in fact, who alerted Jenny Randles to the case. Had this matter not been checked until a long period afterwards, as is usually the case, the *Dynasty* factor would probably have been missed, and the case would no doubt have joined the lists of classic abductions. While the woman clearly underwent a vivid personal episode, it cannot easily be held that an actual, physical abduction did take place.

There is similar uncertainty surrounding reports of structured craft. Given an apparently unusual visual stimulus, such as an unidentified light, some people involuntarily 'see' all kinds of details that are not there. Large-scale meteor or fireball events are sometimes reported as having 'windows'. The stimulus can just as easily be a moon rendered unfamiliar in a hazy cloud, a brilliant planet – or almost anything. This unintentional mental embroidery is called *confabulation*. Some recent work on this has been carried out by Austrian researcher Alex Keul, who holds doctorates in both astronomy and psychology, in conjunction with British teacher and experienced UFO investigator Ken Phillips. They have studied existing psychological and sociological studies of UFO reporters, and have developed an interview scheme called the 'Anamnesis' designed to thoroughly explore the sociological and psychological profiles of UFO reporters. A part of this scheme involves the subject being asked to view a slide of a shape reminiscent of a vaguely structured, luminous night-time 'UFO' for about a minute and then to draw what they saw from memory. These drawings are then subjected to analysis, with certain parameters being statistically evaluated.

In one set of experiments, they showed the slide briefly to two different groups: one composed of amateur astronomers, the other of participants at a ufologists meeting. Most people in both groups tended to get the angle of the object in the 'sky' fairly accurately, but there were substantial discrepancies with shapes and details and in estimated duration of the 'sighting'. About a quarter of the amateur astronomers introduced confabulations into their drawings, while around half the ufologists showed non-existent details in their drawings. These people would naturally have come into contact with more 'UFO folklore' than the astronomers. Keul and Phillips comment on this early research:

> We think we have positively shown that an observer's eyes and mind are no video recorder or computer diskette that the field investigator only needs to 're-wind' to extract exact information, but that people who try to draw uncommon shapes they have seen – especially when the observation was short and details scarce – make great mistakes in time and structures, are better with spatial orientation, and that 'ufological bias' or unconscious channels to UFO folklore material allow a lot of things that were not actually there to materialise in the drawings . . .
>
> The tendency of the human mind to re-structure, omit what has not been seen clearly or put in imaginary details is strong.[3]

This process, it needs to be emphasised, can occur quite unconsciously, and is not a deliberate attempt to deceive. It is also important to note, to avoid the sweeping and unsubstantiated insinuation purveyed by some psychosocial ufologists that there is *no* genuine 'unknown' stimulus for any UFO reports, that these observations relate primarily to details of the shape. Most people recalled something of the basic form they saw.

Research by Keul, Phillips and others in various countries also suggests that a high proportion of UFO reporters claim earlier psychic experiences – a proportion that rises dramatically in 'close encounter' witnesses. UFO witnesses taken as a whole, though, are not some form of 'sub-culture', but display similar sociological characteristics to other members of the community. Specifically close encounter reporters, though, do seem from recent research[4] to form something of a 'social dissatisfaction cluster', with psychological profiles similar to those who claim ESP experiences – a profoundly significant finding as will become clear later on.

There are other factors which could lead to the impression of structured craft. Almost all UFO sightings are of lights, usually seen at night. It seems some earth lights can move together as if choreographed – an interesting characteristic in its own right – and this can readily be interpreted in a night sky as lights on some huge, fantastic dark object. Again, lights of odd and amorphous shapes, or even regular light balls that change colour or have spots of activity on them, could become perfect stimuli for confabulation – the seeing of 'windows' for example.

In daylight, UFOs appear primarily as metallised discs or cloudy, 'soft' cigar shapes. Are these necessarily structured craft? The daylight discs may not be of metal at all: when the Norwegian Institute of Scientific Research and Enlightenment investigated Hessdalen (see Chapter 6) they considered that they could identify conditions that would produce balls of plasma. They made interesting observations on the appearance of such plasma balls in daylight:

> In daytime they often look like metallic balls or discs surrounded by a glow or halo. When such lights are seen they often look as though a metallic object has been placed in the light. Such a description is quite fitting with light from plasma. It is the glowing ionised gas which looks like a metallic fuselage.[5]

A more familiar example to many people of a false metallic appearance caused by one medium suspended in another is the silvery sheen on air bubbles seen under water.

The processes that produce light in these aeroforms also seem capable of producing the complete absence of light, as if light is somehow swallowed up leaving only a patch of absolute blackness – the shape that moved across the road at the Devil's Elbow in Longdendale is one such instance. On the macrocosmic scale, black holes in interstellar space are localised concentrations of collapsed stars producing such intense gravitational fields that even light cannot escape its pull. One can hardly postulate miniature black holes wandering about the terrestrial landscape, perhaps, but the blackness, the sheer lack of light, of some manifestations may simply be the obverse face of the processes producing light in observed earth light or UFO events.

So it can be seen that the small number of accounts of 'structured craft' have to be taken very cautiously indeed. All may not be what it seems, by the time we have taken into account the mysterious properties of genuine anomalous phenomena themselves, and the peculiarities of human perception and psychology. But the lights may interact with witnesses to create further illusions.

FIELDS OF ENCHANTMENT

The ways in which electromagnetic fields possibly surrounding light phenomena could impinge on witnesses have been well considered by Michael Persinger.[6] He has postulated a range of effects, dependent on the observer's distance from the luminous display. His thesis is supported by considerable experimental and clinical evidence which the reader should be aware is largely omitted here. Persinger is not making empty statements.

At a range beyond the influence of the light source's fields, Persinger suggests, a witness will simply, if excitedly, see a curious lightform behaving in a strange manner. As the observer and the light come into closer proximity, details of the lightform will be noted - colours, shape, internal teeming effects and so on. As the light moves closer, or is further approached, the person begins to enter the presumed electromagnetic fields emanating from the luminosity, according to Persinger's scheme. Electrostatic components of the field would initially be registered by the individual as tingling sensations, goose bumps, hair raising, and an oppressive feeling on the chest. As the observer walks further into the field, or as the field moves around the witness, strong currents could be induced within the body, affecting brain tissue and commencing mind-change effects. Particularly electrically sensitive parts of the brain are the temporal lobe areas.

The temporal lobe cortex integrates several sense modes, and is involved with language. Unlike other parts of the brain, the temporal cortex can be rendered electrically unstable or sensitive for prolonged periods, and if sufficiently large parts of the cortex are so effected it is recordable on EEG readings as a focus. Beneath the temporal cortex are two connected structures, the hippocampus and the amygdala. Alterations in the function of the hippocampus can change or modify memory and release dreams into the waking state. The amygdala is associated with emotional feelings. Temporal lobe epilepsy occurs because of the chronic occurrence of tiny electrical seizures within the temporal cortex. A sufferer will report dreamy states, the hearing of voices, the seeing of apparitions and the feeling of compulsions. Experimental, clinical, stimulation of the hippocampus and amygdala produce identical experiences: when small currents are induced in these tissues subjects report scenes or apparitions, experience alterations in time and space or have out-of-the-body experiences. Meaningful messages may be heard that seem to come from the subject's environment. Even brief stimulation can cause hours of alteration in information processing.

There are many stimuli that can effect the temporal lobe, such as electrochemical effects within the brain itself occasioned by the person's behaviour patterns and experiences, the ingesting of certain drugs, or exposure to external electromagnetic fields, as could be the case in the vicinity of earth lights. The mental and physiological conditions obtaining during near-death experiences are also likely to provide appropriate triggers. It seems that the occurrence of electrical disturbance within temporal tissues can be *learned*, and there is no doubt that certain techniques have been developed down the ages by various cultures to promote what we now identify as temporal lobe effects, allowing ESP and mystical experiences to occur. It is even conceivable, as I suggested in *Earth Lights*, that certain classes of ceremonial sites were constructed within known areas of earth light incidence precisely to allow interaction between the phenomena and shamanic representatives of a tribe or group. In the way that primary cultures made great use of local plants and herbs for hallucinogenic (and thus magical and religious) purposes, they would surely have been as aware of the potential of what we today might call geophysical anomalies. The known placing of certain types of megalithic site in northwest Europe in geological areas suitable for earth lights occurrence is quite possibly a result of the use of this environmental method of causing mind-change effects. We noted the 1919 account of lights above the Castlerigg stone circle in Chapter 3, and there are several other known examples of light phenomena being reported at megalithic sites in Britain, and Norse traditions even had special names for lights associated with stone structure sites.[7]

Sequences generated by brief electrical seizures within temporal lobe tissue, initiated by whatever stimulus, can be mild, such as sensations of floating, *déjà-vu* experiences, recurring vivid dreams, memory blanks, feelings of unreality, and so on. Everyone has such feelings at some time or another. More extreme and recurring effects are viewed by our culture as pathological, though in many primary societies they seem to have been given an appropriate social setting and considered signs of mystical, visionary or magical ability. Modern Western culture has profound problems accepting altered states of consciousness, a characteristic that in itself can reach almost pathological levels of intolerance.

Some people are more liable to electrical 'microseizures' within the temporal lobe areas than others, and are consequently more sensitive to appropriate stimuli. This is, no doubt, reflected in the findings of Keul and Phillips where close encounter witnesses exhibit profiles similar to ESP percipients. It is possible, therefore, that such a person entering electromagnetic fields associated with earth light appearance might be 'triggered' more readily than someone less electrically sensitive.

As the close encounter with the light phenomenon continues, the witness would experience modifications of consciousness, vision and memory. Emotional states, such as terror, or, conversely, religious awe, could also be generated. If electrical aspects of the field dominated, then localised effects on neck and thighs would occur, whereas if magnetic components were dominant, then waist and genital regions would be especially affected. Sensations in these parts of the body could readily be incorporated into the mental states being created within the witness by the external fields. (This happens normally to people in sleep: a physical signal such as thirst can enter a sleeper's awareness as a dream about wandering in a parched desert, or a dream of running water might be the body suggesting that a quick trip to the bathroom is called for!) As the person and the light draw yet closer to one another, the witness might undergo a seizure with ensuing unconsciousness. On awakening, there may be partial or total amnesia. Persinger considers that, dependent on the size of the light and the amount of energy sustaining it, even closer contact could result in death, or physical effects such as burns, hair loss, forms of radiation sickness and the like. (Such energetic lightforms could also leave physical traces in the environment, such as burnt or dried patches of ground or singed foliage.) It certainly seems to be the case that a 'spirit-eater' light figures in the lore of different Indian tribes along the San Andreas Fault, for example, which indicates dire consequences for anyone unfortunate enough to come into contact with such a luminosity, and close encounter with the 'chota-admis' lights of Darjeeling (Chapter 6) were considered likely to cause illness or death. Nevertheless, there are also numerous accounts of people coming very close to lights without receiving any physical injury, as well as being close enough to see objective details without any apparent intrusion of phantasmagoria. It may all be a question of at what frequency the energy from a light emanates.

Persinger points out that 'post-stimulation electrical instability . . . allows the creation of transient neuronal firing patterns that do not necessarily represent concurrent sensory input. Instead, they could incorporate combinations of memory and fantasy, but still within the context of the moment.'[8] Memories both of the event itself, and of time before and after it, could thus be modified by such continuing instability.

This all has the profoundest implications for claimed abductions where it can be reasonably assumed that *something* did actually happen. All abductees are not self-generating their experiences psychologically: sometimes they are pushed into them by external phenomena.

Let us take an important abduction case, important in the sense that Bud Hopkins bases his book *Intruders*[9] around it. Hopkins has been a catalyst in the USA for the emergence of what can only be called abduction mania. To those researchers outside the American ufological *milieu*, this lemming-like stampede towards the assumption that people by the hundreds are being abducted and interfered with by aliens generates a feeling similar to that which might have been felt by a stranger passing through Salem at the height of the witchcraft hysteria.

In *Intruders*, Hopkins describes the case of a key subject he calls Kathie Davis. As a result of the memory of an incident and subsequent disturbing dreams and feelings, Kathie contacted Hopkins on account of an earlier book of his (*Missing Time*) which dealt with UFO abductions and which she felt resonated with her confused feelings. The bare bones of the situation are that Kathie had experienced a close encounter in her back garden in Indiana, lost at least an hour of memory, and suffered certain mental and physical symptoms. Hopkins had Kathie undergo a series of hypnotic regressions which, mosaic-like, built up a story of abduction by apparent extra-terrestrials, at the hands of whom Kathie was subjected to highly invasive examinations. It appears she also was inseminated, and a half-human, half-alien child resulted from this. According to the hypnotically induced memories, this offspring was in the care of the extra-terrestrials. Hopkins brings in accounts from local witnesses who had experiences that seem to suggest something odd did indeed happen, and he describes other people who feel similar experiences happened to them. The core, trigger experience is soon forgotten in the book amidst the welter of images derived from the hypnosis and constructs placed upon them by the author and others.

What seems to have happened is that Kathie came close to a *ball of light* in her back garden. She first noted a light in the swimming pool house, and brought her mother's attention to it, thinking she had accidentally left the light on there. When Kathie checked a short while later, there was no light and everything seemed in order. She visited a neighbour. During her absence, however, Mary, Kathie's mother, saw a light this time alongside the pool house. She could see the bird-feeder through it. The light was curious, 'there wasn't any beam', and it was 'about as big as a basketball'. Then it 'sort of faded out, all at once.' She called the house where Kathy was, as she had an uneasy feeling. Kathie returned and went out into the garden with an (unloaded) gun to investigate. This was at 9.30 p.m. She recalled being outside for only ten minutes, finding nothing unusual. She then went to a friend's house to suggest a moonlight swim in the pool, as it was such a hot night. She should have arrived at her friend's place no later than 10.00 p.m. if only ten minutes had elapsed in the garden. In fact it was 11.00 p.m. when she got there. Kathie, her friend and another companion returned to the Davis's home and entered the swimming pool. They all began to experience odd sensations such as irrational coldness, either a fogging of vision or a highly localised haze condition in which 'haloes' could be seen around the electric lights. All three swimmers felt nauseous at about the same time.

Hopkins discovered that the Davis's closest neighbours had noticed some unusual occurrences that evening. They had seen a sudden flash in the direction of the Davis's house and felt a low, vibrating sound. They felt their own house shake and a chandelier moved slightly. The TV set displayed interference, the house lights dimmed and the digital clocks in the house had to be reset. These neighbours felt there had been a small earth tremor. The next morning the Davis's lawn showed a burnt circle with a long line leading to it.

All this was interpreted by Hopkins as evidence of a UFO landing, but the reader of this book will be aware that something else, by now quite familiar, had happened. Energy release over a few hours had occurred as the result of seismic stress. Electromagnetic fields, light phenomena and tremor activity were produced at various points in the episode. Standard 'basket ball' earth lights had appeared, and one of them had burnt the grass, marking its passage close to the ground for several yards then the spot where it hovered before vanishing. Kathie came in close contact with one of these lights, and experienced the precise sequence of effects described above in Persinger's scheme. Her mother also experienced similar but less pronounced effects, probably due to distance. Both women were within the sphere of influence of the lights' fields, and probably the ambient fields that helped produce the lights themselves. These latter were still present by the time of the moonlight swim, causing typical symptoms – even localised fogs are a frequently reported earthquake effect, though in the Davis case physiological symptoms may have been responsible for the effects on vision.

The hypnotic regressions probed the period of amnesia and the peripheral periods of confused memory and probably merely elicited the content of imagery encoded in neuron patterns within the temporal lobe area. In Kathie's hypnotically accessed memory, we see a hybrid of traditional fairy changeling themes amalgamated with modern UFO folklore, spiced with psychologically induced imagery relating to the woman's response to living in a modern, technological society with all its pressures, fears and dehumanising aspects.

In my view, Kathie experienced remarkable geophysical and psychological effects that are instructive to us all, rather than undergoing an extra-terrestrial visitation.

Whitley Strieber, too, has been a key catalyst in the late 1980s abduction craze with his best-selling *Communion*,[10] in which he gives an apparently truthful account, derived from fragmentary memories, dreams and hypnotic regression, of being 'floated' out of a bedroom by 'Visitors' to an unusual craft or environment and undergoing intrusive examinations. Strieber is wisely cautious about automatically assuming that these Visitors are extra-terrestrial. He comments: 'Maybe we have a relationship with our own planet that we do not understand at all'. But while his intellectual self wondered what the nature of the Visitors could be, his emotional self 'did not share the indecision'. He *felt* that he had encountered real people, though non-human ones. Of course, such feelings accompany stimulated temporal cortex sensations too. It is extraordinarily interesting that at the time of this writing (July 1988) it is reported[11] that Strieber has undergone complex neurological examinations with a Magnetic Resolution Scanner in New York. 'Three punctate foci' were apparently detected in 'the cerebal white matter of the frontal lobe and the temporal parietal regions' of Strieber's brain. Strieber seems very healthy, mentally and physically. In his recalled experience he felt that needles were inserted into his brain by the Visitors. Perhaps the neurological examination is providing evidence for inserts made in his brain by the Visitors, but perhaps his recalled imagery is related to natural abnormalities (not necessarily harmful) occurring within his brain's cortex, and which may themselves have caused his extraordinary experiences.

It is certainly the case that many abductees report floating to the alien 'craft', floating being a typical sensation produced by temporal lobe stimulation. In near-death experiences, people feel themselves floating along or up a tunnel towards a light. Abductees sometimes experience being drawn up a lightbeam towards the 'craft'. The imagery and sense of motion are very similar. There are many facets shared by abduction, out-of-the-body and near-death experiences. It is similarity that should not go unnoticed.

Jenny Randles has noted a 'kind of sensory isolation' that is a 'very common feature' in reports made by close encounter witnesses. They feel a strange, unreal mood settling over them at the onset of the event. Sounds diminish, the environment seems unusually quiet and devoid of activity. Randles calls this the 'Oz-factor'.[12] In fact, such an *abaissement du niveau mental* is a state which C. G. Jung considered to be 'a very important pre-condition for the occurrence of spontaneous psychic phenomena' and which the great psychologist linked to close encounter UFO experiences as long ago as 1958.[13]

THE POLTERGEIST EFFECT

Another aspect of earth light events the reader will have noticed in Part 2, is how often there is reference to poltergeist or other paranormal activity occurring in the immediate neighbourhood of earth light outbreaks, and over the same period.

Persinger extends his thesis involving disturbance of the temporal lobe areas of the brain to account for this. He suggests that transient geophysical fields are produced in areas of tectonic stress which not only can manifest as light phenomena but can affect human perception and, in a weaker form, cause effects that are associated with poltergeist activity.[14] This could simply include events like brief electrical failures (as in the Kathie Davis case) and effects on radio and TV reception. Strange apparitions and intense feelings of dread, terror and so on could be generated by electrical disturbance of the temporal lobe. Non-conductive objects – glass, crockery and the like – could be displaced, their motion or spin reflecting the vector (magnitude and direction) of the field. These tectonically-generated fields would probably be accompanied by brief alterations in gravity, Persinger argues, but while large-scale correlations between gravity and magnetic anomalies have been scientifically studied, there has been no work on highly localised, brief events. Persinger expects physical effects within a house to show a relationship with the layout of conducting parts of a house, such as the plumbing and wiring. The effect on some people of brief but intense geophysical fields would be to interfere with cardiac or respiratory activity creating a sense of being suffocated. Many of these conditions have been widely reported by poltergeist investigators, though Persinger is at pains to point out he is not attempting to explain all poltergeist phenomena.

In 1975 Persinger was himself called to an apartment in Sudbury, Ontario, which had become suddenly haunted by strange sounds and fleeting images, with the two female occupants frequently feeling fear and terror.[15] Indeed, one of the women sometimes 'felt' apparitional images instead of actually seeing them. Although Persinger was able to arrange some electronic monitoring at the focus of the events within the apartment only towards the end of the active period, on two occasions the instruments recorded 'electromagnetic-like' vibrations, one of them particularly severe, involving a spill of ink on the pen recorder chart.

The paranormal events in the apartment had increased in intensity and frequency prior to an outbreak of light phenomena in the immediate Sudbury area – the greatest incidence of UFOs ever reported there. The period of the UFOs and the poltergeist activity overlapped. It was also found that the small apartment block was adjacent to an outcrop of rock, there was local mineralisation, including copper, a minor fault ran a few hundred metres from the apartment and a major fault was also close by. Years after the UFO flap, checking by Persinger with the apartment block owners indicated that the poltergeist events had ceased or diminished to very infrequent occurrence.

Something akin to Persinger's ideas is well supported by a poltergeist event on the Isle of Lewis, off Scotland's western coast and the scene of earth light phenomena (Chapter 3). It was well understood at the time (not given, but was investigated by the Society for Psychical Research) by journalist James Shaw Grant.[16] The incident occurred at the village of Tolsta Chaolais in 'a little timber house perched on a spur of rock' one Sunday morning. The house was inhabited by an old lady who was entertaining her two grandchildren. In full view of the woman caorans (small peats) began to jump from beside the 'very large cast-iron cooking stove'. Then the glass chimney of a hanging lamp crashed to the floor. This was followed by a jug sailing through the air and the mass breakage of crockery. The events caused so much noice that other members of the woman's family nearby came rushing in to see what was going on. They found a row of cup handles still hanging on their hooks, but the cups themselves had fallen to the floor; the elderly woman's toothbrush was snapped; plates in the kitchen sink were cracked clean across, and a bar of soap was cut into three pieces. Grant questioned the old woman the next day. He found her to have 'a clear mind and remarkable strength of character' and he had no doubt about her sincerity. The woman said she did not understand what had happened 'but I know there is an explanation'. Grant felt he knew what it was. He noted that the poltergeist incident had occurred 'during a period of quite unusual electrical activity all over Europe'. It was a weekend 'characterised by briliant displays of aurora borealis at night, and violent thunderstorms by day.' The night before the poltergeist incident, Grant and a friend has stood out of doors watching the northern lights. Instead of the familiar glowing green curtain of light, there were 'great gouts of pink and crimson light dancing all around us, from the horizon right to the zenith.' Grant further reasoned:

> Another feature which points to an electrical disturbance was the structure of the house . . . (it) was a perfect Leyden jar, in which the charge would be built up. . . . The wooden house provided the non-conducting container. Inside it was a disproportionately large iron stove, with an iron chimney projecting through the roof. The first indication that something unusual was happening came when the caorans began to jump away from the stove, as the room became charged with electricity via the chimney . . . .
>
> A few months after the incident the old lady had occasion to darn a sock. She used a ball of wool, which she had found, on the day of the poltergeist (which was not a poltergeist). . . .But when she tried to darn a sock with it, the wool crumbled in her hand!
>
> The significance of that was pointed by an incident in West Uig not long afterwards. A crofter, out working on his fence, was struck by lightning. . . . He was uninjured, but his trousers were ripped off. Some time later is was discovered that all the clothes kept in a trunk in the house had mysteriously perished, just like the ball of wool.

Tolsta Chaolais is on the eastern side of East Loch Roag, adjacent to the important local fault which runs along there, and about 2 miles (3 km) from the scene of the Loch Carloway light mentioned in Chapter 3.

Livingstone Gearheart and Michael Persinger have found strong statistical evidence amongst poltergeist cases in North America and Europe to support the view that poltergeist activity in general seems to occur by preference at times of significantly high global geomagnetic activity.[17] On the other hand, Persinger and colleagues have produced stunningly convincing statistics from excellent data bases to suggest that intense telepathic and clairvoyant experiences occur on days of *low* global geomagnetic activity.[18, 19, 20] Persinger and other researchers such as Puthoff and Targ have suggested that spontaneous experiences of ESP may be associated with Extra Low Frequency (ELF) electromagnetic fields, as these have very long wavelengths that travel long distances, and can penetrate buildings and water. A major source of natural ELF fields are the waves generated in the space or cavity between the Earth's surface and the ionosphere, known as the Schumann resonances. These fields pulsate at rhythms similar to some brain waves.[21]

We have seen that Helmut Tributsch's idea of charged particles generated by rocks during tectonic stress has been experimentally confirmed by Stuart Hoenig. In *When the Snakes Awake* Tributsch presents evidence to show that airborne ions can have 'a pervasive effect on the contentration of serotonin . . . a nerve hormone that abounds in the lower middle brain. Serotonin can cause far-reaching changes in the body's metabolic processes, and also affects the transmissions of nerve impulses, sleep, and the development of moods.'

Taking all this information together, we can see that we are in an extremely complex area when considering abductions and the sighting of apparent structured, metallic craft. We have entered a jungle of unfamiliar geophysics, sociology, psychology, perception and neurophysiology. We can scarcely begin to calculate how all these various mechanisms, and others we may be quite ignorant of, merge together at certain times and places and what effect they can have when they do so. Clearly, we do indeed share with our planet a complex relationship we have as yet hardly any knowledge of, as Strieber tentatively suggested.

Let us take one more step towards tomorrow's science. If earth lights and their ambient fields and supportive conditions can affect us, can we have an effect on them? The answer, I suspect, is 'yes'.

A FORBIDDEN VISION

There is evidence in sighting reports that suggests the lights can respond to witness movement and *thought* – that they are consciousness-sensitive. I first drew attention to this possibility in 1982 in *Earth Lights*. It was either ridiculed or politely ignored. In bringing it up again, I realise I am courting summary outrage or dismissal from almost everyone. I confess that at times I feel almost ready to apologise for putting such an idea forward. The reason I am prepared to persist in this apparent foolishness, to stand virtually alone on it, is primarily due to my own personal experience and the statements made by many other witnesses around the world. The inability to entertain such a possibility may reflect psychological discomfort on the part of sceptics rather than what actually happens with light phenomena: a subjective rather than objective response.

My experience of 1967, at Bromley, Kent, was described in detail in *Earth Lights*, so I will do no more than outline it here. In May of that year I saw from the top floor of what was then Ravensbourne College of Art and Technology a brilliant orange upright rectangle of light moving through the sky from the north, the direction of Bromley, over Bromley Common to the general vicinity of the college. The light was pulsing, and had a perfect outline, in the proportions of a door. There were other witnesses in the same room as myself, and more came out onto the car park below. The light came to a standstill some hundreds of feet above fields about a quarter to half a mile from the college. The light intensity subsided a little, then the regular outline of the lightform collapsed in on itself and, while maintaining a stationary position in the air, it proceeded to boil and billow like an animated cloud which was by now glowing with the colour of embers. No one in the room I was in could move a muscle: we were transfixed. Then something happened that created a sense of awe that is impossible to communicate adequately. The boiling cloud suddenly restructured itself into another clear form, the animated inner motion ceasing. I saw the featureless, glowing orange shape of a figure with its arms outspread. Perhaps because of my Catholic upbringing I associated this with an angel or a Christ figure. It turned out that some witnesses saw something like Vitruvius' or Leonardo da Vinci's 'Universal Man' figure while others saw a simple cross shape. I suspect there was a basic form that each person perceived through the filters of their own psychology and background – it was clear that everyone saw something very similar. (Later that year people in various parts of Britain reported 'flying crosses'.) I was so dumbstruck that I cannot now recall the exact period of time this shape remained in the sky. But after perhaps a minute or two, this shape also collapsed in on itself. This time the activity in the cloud subsided and the light became duller. Fully fifteen minutes later, there was still a rosy smudge in the sky where the phenomenon had taken place, gradually dissipating as if on air currents.

To anyone who did not see this phenomenon, the above account has no objective value. I wish I could scoop it out of my memory and place it on a table for all to see and accept as real. It did actually happen, and those of us who observed the event have to cope with the objective reality of that fact in our various ways. Non-witnesses have the subjective luxury of either believing or disbelieving. In an attempt to bring some additional evidence to bear, I suggested in *Earth Lights* that a sceptic qualified to conduct a serious investigation might consider researching this incident, chasing up as many other witnesses as were now traceable and conducting interviews with everyone, checking the site of the event, and carrying out polygraph (lie detector) tests – I for one would be open to this. Polygraph tests would not add up to much, however, unless more than one witness agreed to them. If this could be achieved, the actuality of the event I have described could be better established. I was slightly disappointed to find that no one took up this suggestion. It would take time and some expense on someone's part, but, then, so does all research.

The event happened, but was I victim of temporal lobe phantasmagoria induced by energy fields from the orange light? I do not think so. For one thing, the light phenomenon was a fair distance away and many of its witnesses, including me, were inside a modern, steel-framed building. Second, everyone I spoke to saw the change of shapes I perceived, even if there were slight variations in interpretation of that shape. (Such small variations would occur in multi-witness reports of any objective event.) Third, after the main part of the incident, there remained the rosy smudge in the sky, attesting to some objective material having been present. I feel the only valid observation is that the lightform was able to change shape – a characteristic long noted with earth lights. The crux is, however, that in this case two regular, coherent shapes, a rectangle and a cross/figure, were formed. These two shapes can be conceptually linked, as I consciously realised only three years later, when the geometer and architectural researcher Keith Critchlow mentioned during a lecture that the door was basically a root five proportional rectangle, because it was designed to be used by another root five proportional system, the human frame! The Bromley light had looked like a glowing door. The 'figure' was seen by some witnesses as a depiction of the proportional human form. *The forms of the Bromley light therefore manifested an idea.* The only source I know of for ideas is the human mind – either the individual consciousness or some autonomous, collective level of mind.

It was this experience that convinced me that the energy comprising earth lights can be sensitive to the input of consciousness. What I saw was not a spacecraft, but it did display intelligence in the form of a visually transmitted concept.

Earth light phenomena mostly hold to well defined, usually geometrical, shapes. Sometimes, though, they shape-shift, but usually from one amorphous form to another. But just occasionally, vaguely human-like, anthropomorphic, shapes are reportedly glimpsed within the light energy ('occupants'), and, sometimes, the lights themselves can take on such forms. I call these latter shapes 'proto-entities'. In the Barmouth outbreak of 1904–5, a distinct proto-entity was reported. Local journalist Beriah Evans stated in a despatch to the *Guardian* (16 February 1905) that a strange light effect was witnessed near Bryncrug, a village to the south of the main Barmouth–Egryn area:

> A professional gentleman returning homeward suddenly saw a gigantic human figure rising over the hedgerow, with the right arm extended over the road. Than a ball of fire appeared above, and a long white ray, descending, pierced the figure, which vanished. Only this long ray was white; the other shorter rays were blue. This extraordinary manifestation – apparition and light – was witnessed simultaneously by a prominent local farmer from another standpoint. The same light again appeared to both.

A case from Cumbria involved a group of young people who were in a pony and trap one moonlit night on their way to climb the Cumbrian peak of Skiddaw:

> While passing Bassenthwaite a hush suddenly fell upon the party for they all saw an apparition rising from the hedge. It rose until it reached human height, its deathly whiteness standing out against the darkness of the hillside, and then slowly made its way to the top of the hill where it stood a shining spectre.[22]

This case is obviously being described from a supernatural perspective, but we can see quite clearly that it sounds like the vaguely anthropomorphic 'white lady' type of vaporous luminescent column. Because these things are usually reported in the context of ghosts and ghost stories, they have not been readily associated with lights and associated phenomena. They are almost certainly related, however, as the following examples will illustrate.

Both luminous and shadowy phenomena are seen periodically on Kings Sedgemoor, beneath Polden Hill in Somerset. The cultural interpretation is that they are ghosts of soldiers from a seventeenth-century battle that took place at the site. What is *actually* seen are 'shadowy figures, almost clouds of vapour' by day while 'at night the apparitions are of blobs of light which expand and contract in size while maintaining a steady green glow.'[23] In 1928 at Chesterton, near Stamford in Lincolnshire, a 'monk-like' figure was seen to emerge from a swamp. It was said to have been luminous 'like the dial of a watch'. It, or a similar phenomenon, was also seen hovering by an iron bridge nearby.[24] In recent years shimmering white or blue-white 'figures' have been reported at Spofforth Castle, Wetherby, Yorkshire; at the village of Oxenden south of the old county of Rutland; on a country lane near Highworth, Wiltshire, and other locations. One night in February 1970, two men driving into Colchester, Essex (scene of the greatest recorded UK earthquakes – see below) saw a shimmering white mist 'like two lines standing up in the road', which they drove through. In this case the 'white lady', 'monk' or other ghostly association was not made. In a discussion on Will-o'-the-Wisp, a nineteenth-century case was recorded where 'a gentleman, a native of Shropshire, whose veracity is above suspicion, viewing a blazing sphere through a spy-glass, discovered in it the lineaments of the human form.'[25] A mound known as Castle Hill at Newton-le-Willows in Lancashire is the locale of a 'white mass' or 'a white shape . . . nearly six feet tall'. This has been reported a great many times, usually on a nearby road. It flits or glides, and sometimes stays stationary. It is variously interpreted as a 'white lady' or 'a monk', and some observers claimed to see some details of a figure within the light. Its appearance has caused at least one traffic accident.[26,27]

Borley Rectory in Essex is considered one of Britain's most haunted spots, usually associated with poltergeist activity. This 'ghostly' label unfortunately obscures a situation which the reader must by now be finding quite obvious. Many investigators at Borley have reported *light phenomena*. Lights have been seen in the branches of trees surrounding the place, and 'pin-points of light that hovered over the Rectory garden at a height of ten to fifteen feet' have been observed. Electronic failures (cameras, car electrical systems and so on) are fairly common, and voices and disembodied footsteps have been heard, as at many earth lights locations (Yakima, for instance). Also the vague luminous figure: one sighting involved the appearance of a 'luminous white figure' in the churchyard that changed to 'a luminous white patch' that moved around slowly. This incident was accompanied by car failure. This phenomenon is seen periodically, and is usually perceived as a woman in a long white gown, invariably considered to be a 'nun' as a legend relates to the horrific death of a nun centuries ago at the site, which was previously occupied by a Benedictine abbey. Borley is in an area with a history of exceptional (for Britain) seismicity, being only about 18 miles (29 km) from Colchester which experienced the great earthquake of 1884. If it has not yet been done, the Borley site needs a thorough seismological and geophysical study.

The ghost literature is replete with such accounts. The columnar phenomena involved in these reports may simply be interpreted as figures by onlookers – certainly the more detailed assumptions are likely to be due to confabulation – but the cumulutive effect of witnesses' descriptions is to create the picture of a vaguely human-shaped form that behaves as if possessing some limited awareness of its surroundings. Does this luminous vapour-column, so often associated with outbreaks of earth lights (see the Linley case in Chapter 3, for example), mould itself to some extent in response to the assumptions or expectations placed on it by almost every witness that encounters it? After the Bromley experience, I am inclined to think this is possible.

Light phenomena take on other shapes as well. I was impressed in the Yakima reservation accounts (Chapter 5) how some lights seemed to mimic tree fires, an image certainly in the minds of the fire lookouts, and how some of the Hessdalen lights (Chapter 6) took on the appearance of 'Christmas trees' – a most Nordic motif!

It should be noted in passing that the wayward electrical genius Nikola Tesla also felt that mental imagery could be sustained outside of the human brain: 'it should be possible to project on a screen the image of any object one conceives and make it visible. Such an advance would revolutionise all human reactions. I am convinced this wonder can and will be accomplished in time to come.'[28] He was foreseeing an electrical invention of some kind – and he did develop a most sensitive electrical discharge – rather than the use of a natural phenomenon.

Another, and perhaps the most pertinent manner in which earth lights seem to exhibit a display of intelligence is the way they appear to be aware of the presence of observers, sometimes playing with them or teasing them. There are numerous examples given in Part 2 of this behaviour. Caution in interpretation has to be exercised, however. The movement of some lights could simply be in response to human biofields, the approach of a metallic car, and so on. A ball bearing rolling towards a magnet would seem to possess rudimentary intelligence to someone ignorant of magnetism. It is all too easy to read meaningful behaviour into random movement as John Derr points out:

> Marsha Adams and I have seen people convince each other that lights were responding to their flashlight signals, when the real source of the lights, as we determined by aerial reconnaissance the next day, was almost certainly headlights on a vehicle descending a slope some 20 or 30 miles distant. The variation in light intensity was caused by the motion of the vehicle winding its way down the rough road.[29]

The matter is obviously fraught with difficulty, but I do think one has to take at least some notice of cases of apparent intelligent behaviour referred to in Part 2, and elsewhere in the literature. To explain away by the above means how the Marfa lights interacted with the two geologists (Chapter 5), for example, seems to be stretching things a little. Another case that would be equally difficult to dismiss on the grounds of coincidence took place at Holker, Cumbria, in 1817, when James Stockdale, the brother of a local historian, saw 'a pale phosphoric light, rather bright but not flashing or sparkling, the size and shape of a pineapple' gleaming on a wall. As Stockdale approached, the light flickered down to the road, and did so every time he attempted an approach. When the man walked off, the light followed him. 'When he stopped, the light hopped on the wall again, until he reached High Row and home.'[30]

Biologist Frank Salisbury made special note of the apparent reaction of light phenomena to witnesses during the Uintah Basin flap (Chapter 5). 'Many witnesses reported the feeling that the UFOs seemed to *react* to their actions or even their thoughts,' he wrote. 'Reaction to *thoughts* would be difficult to prove . . . . Nevertheless, there are many cases in which some action taken by the observer . . . resulted in rapid departure of the UFO. This seems to be a clearly discernible pattern.'[31]

The leaders of the only two field studies undertaken while light phenomena were actually occurring – at Hessdalen and Piedmont, Missouri – are themselves sure that the behaviour of some of the lights they dealt with were actual responses to observers. Members of the Hessdalen team told me this directly, and Leif Havik has written that he no longer carries his camera when walking around Hessdalen, 'the reason for that is that when I had my earlier sightings (approximately 30–40), I have noticed that the phenomenon in a few cases has disappeared into nothing when I was trying to photograph it. Therefore I have been testing to walk around in the mountains without any camera or other instruments, to see if it might be affecting the appearance of the phenomenon.'[32] A similar reaction was noted by witnesses to the Hudson Valley phenomena (Chapter 5):

people who attempted to use flash cameras in taking pictures of the lights, found the phenomena would start to move towards them – something that did not happen to those who used cameras without flash guns. This found an extraordinary echo in something police sergeant Tony Dodd told me regarding his experience with lights on the Yorkshire moors (Chapter 4). He mentioned that the lights seemed to be attracted to the flashing beacon on top of his police car – an observation noted by some of his colleagues, too. Dodd is quite convinced that the lights display intelligence. He has seen them dart behind trees or hills when an aircraft approaches, and they have played games of 'tag' with him and other witnesses at times – markedly similar to behaviour described by the Marfa geologists.

Dr Harley Rutledge, who led the Piedmont flap team (Chapter 5), came away from the field study convinced that there was intelligence involved in at least some of the lights' behaviour patterns;

> Thirty-two Project cases of apparent reaction or awareness have been counted. . . . On the second night I was at Piedmont, experiences suggested to me that . . . the UFOs were aware of our presence, that the UFOs may have purposely attracted our attention, and that they may have reacted to us – although at the time I did not label the sightings as UFOs . . . .
>
> How did the UFOs react to us? They turned lights off, on, moved away, shot away, changed course, changed brightness, and the like . . . .
>
> A relationship, a cognisance, between us and the UFO intelligence evolved. A game was played. In my opinion, this additional consideration is more important than the measurements or establishing that the phenomenon exists. This facet of the UFO phenomenon perturbed me. . . . It is an aspect I cannot really fathom – and I have thought about it every day for more than seven years.[33]

Rutledge also felt that the lights 'do imitate natural and man-made phenomena'.

While it would be rash to state that earth lights definitely do display intelligence, I think it would be most unwise to dismiss the indications we possess that such could be the case. Certainly the phenomenon I saw at Bromley *did* reveal a sentient component, and it could hardly be a unique occurrence.

If we accept earth light intelligence, at least for the sake of argument, where can it come from? Telepathic pilots aboard balls of light sometimes only inches across? Telepathic controllers on mother ships controlling plasma-drones by remote control? I do not think so. It could be that the lights are some form of life. Perhaps they are spirits, as many early and primary peoples claimed. But this is unreliable, because unusual phenomena were habitually classed as spirits by such people. In any event, what are spirits? I continue to feel that we are dealing with a very sensitive energy form, sufficiently sensitive to react to consciousness itself. People today may have difficulty with such a proposal because our concepts of what consciousness is could be limiting. There is a strong tendency to think of consciousness as resulting purely from complex interactions within the human brain, and that when the brain dies, consciousness ceases. This is the mechanistic view, which could be entirely mistaken. While there is evidence, as we have seen earlier in this chapter alone, that parts of the brain can be identified with certain functions of consciousness, this is not the same as saying it produces consciousness. If we can identify brain regions that are involved with the function of memory, for instance, it is not the same as understanding or locating memory itself. The brain is more likely to be a wonderfully complex *processing organ*. Lungs process air, but do not create air; eyes process light, but do not create light. In my opinion, we need to start thinking of consciousness as a *field effect*, an all-pervading element in the universe, perhaps associated with space-time in ways not currently apparent to us, and affected by the presence of electromagnetism and mass. Such a field would allow the operation of ESP phenomena to be construed in fresh ways.

If this view is correct, an earth light and its witness would both be immersed in the consciousness field. In the way that a person at one side of a pond can make a cork in the middle move up and down without directly touching it, simply by putting energy into the medium, the water, in which it floats, so the processing of consciousness within a witness' brain may have effects on the consciousness field that in turn affects the light phenomenon. This process could be augmented by electromagnetic influences on the observer's brain function and on the consciousness field itself.

But, in the final analysis, the nature of the lights will automatically be revealed if we continue to study them, and progress in that effort. There is scientific scepticism – but a few scientists, as we have seen, persist in studying the lights. There are sceptics also within ufology, but denial is not the same as objectivity. In my view, scepticism about the actual existence of earth lights is no longer a reasonable option on anyone's part. The evidence worldwide is enormous. However unsatisfactory that evidence may be for some tastes, it is too widespread and too homogenous to be dismissed by anyone pretending to a scientific attitude. No subject, let alone that of terrain-related lights, could ever progress if everyone in the world had to wait for what they would accept as incontrovertable proof. There comes a time when enough evidence accumulates to require the emergence of a consensus opinion that a phenomenon actually exists. That time has arrived with earth lights. The response should be to go out and improve the quality of the evidence, not to ignore it. By the use of this book alone, locations can now be singled out where the likelihood of seeing light phenomena is greatly enhanced. We are actually beginning to learn the geography of the UFO, of the earth light. Some ufologists do now accept earth lights as a natural phenomenon, but make a distinction between them and UFOs, as though UFOs were some kind of identified quantity. One is entitled to ask: where, then, do earth lights end and UFOs begin? Then there are ufologists who dismiss earth lights because they fear a challenge to their pet theses. But I have already explained that even ETH adherents may have something to gain by a study of terrain-related lights. Whoever we are, we can have our own theories - and sceptical opinions have no automatic superiority over other kinds - but they should not be allowed to repress or hinder research.

If some of the ideas discussed in this chapter seem far-fetched, then it is because the subject we are trying to understand is far-fetched. While parts of the earth lights enigma almost certainly intrude into areas accessible to our current levels of knowledge, other aspects stretch out beyond the present reach of our understanding. That will be true whatever the final explanation for the lights is. The earth lights thesis does *not* dismiss light phenomena simply as 'balls of electricity'. They are either an entirely unknown phenomenon, or an exotic development of known forces. Their study is going to lead us to the comprehension of a hitherto unknown form of energy, which may have unguessed properties and applications (think of communicating with a computer via a consciousness-sensitive interface, for example!). But because unravelling the earth lights riddle will demand radically different appreciation of our relationship with Earth and the nature of consciousness, it may be some time before we can evolve sufficiently to even ask the right questions. Perhaps we never will, but the more optimistic, who wish to be in 'on the ground floor' of what could be the energy base for the next millennium, should be beating a path to the doors of those researching terrain-related lights.

Indeed, it is difficult to believe that some governments, at least, are not aware of the existence of earth lights, and their potential value. ETH *afficiandos* often claim government conspiracies involved in keeping secret the whereabouts of crashed UFOs and dead occupants, and see themselves as intrepid investigators stripping away the layers of governmental skulduggery. It is hard to understand how anyone can think like this today. Are we to believe, as some ETH proponents suggest, that America has had UFOs and dead aliens locked away for 40 or more years? Does anyone think that such a secret could have been kept, when other, much more recent top-secret information leaks like water from a sieve? The number of people - military and scientific - who would have to have been involved with such enormously significant material would be prodigious. (Even the lunar material brought back by the Apollo missions was passed to laboratories all over the world for analysis.) Moreover, where is the evidence that anything has been gleaned from this advanced ET technology? We can trace the lineage of terrestrial hi-tech developments. It is ours, together with all its shortcomings: space shuttles still tragically blow up. And where is the hint that the US government acts in the certain knowledge that ET civilisations exist, and that they have contacted Earth? Where is the change of behaviour pattern that would be bound to show from such a tremendous psychological shift?

If a conspiracy exists, I suspect it is too subtle for the gung-ho ETH brigade. What if – I put it no stronger – governments deliberately leak phoney 'secret' documents every so often? Just enough to convince the ETH adherents that they are on the right path, steering them well away from other research on UFOs which might lead to real findings officialdom would prefer were kept secret. At the time of this writing, the abduction craze has been accompanied by the 'uncovering' of secret documents relating to MJ-12 – Majestic 12, a supposedly secret group of former officials in America aware of the US possession of a crashed 'saucer' and occupants. If it is a governmental ruse, it is working beautifully: many ufologists, especially American ones, take it all very seriously. American ufology, as a whole, seems to have been successfully deflected from other lines of enquiry. The plot, if such it be, is not only influencing independent ufological research, it is affecting what publishers see as viable material. A stranglehold is thus placed on thought and communication. There could be a master psychologist at work in the Pentagon employing the crude but effect maxim – 'Give 'em what they want!'

But all that is for devotees of conspiracies. It passes the time. For those who do not care whether such scenarios exist or not, for those who do not have time to waste, who want to engage on serious UFO research, there are more fruitful avenues to explore. The study of earth lights is one of them. Their pursuit will lead us to a deeper understanding of consciousness as well as of geophysics. The phenomenon has in its power the ability to perform a kind of alchemy in which mind, electromagnetism and the terrestrial environment can be brought together in a new conceptual framework. Earth lights research holds the potential of creating a whole new area of human study, one of evolutionary significance, one that can help heal the fragmentation that bedevils our contemporary way of thinking about so many matters.

In at least a symbolic sense, earth lights may be leading us back to ourselves, and back to an exciting new realisation of the true nature of our relationship with Earth – the world we have for so long now ignored, damaged and undervalued. The lights can show us the way back home.

REFERENCES

CHAPTER 1 EARTH'S LIGHT SHOW

1 Blesson, L., *Entomological Magazine*, 1833.
2 Quoted by A.A. Mills, *Chemistry in Britain*, February 1980.
3 Reeder, Phil, *Northern Earth Mysteries*, 30, 1986.
4 Mills, A.A., 1980, *op cit.*
5 Leeson, C., *Notes and Queries*, 4 April 1891.
6 Erskine, James M. Monteith, *Daily Mail*, 3 January 1913.
7 Derr, John, *Bulletin of the Seismological Society of America*, 63, 1973.
8 Fuller, Myron L., 'The New Madrid Earthquake', *US Geographical Survey Bulletin*, 494, 1912.
9 Tributsch, Helmut, *When the Snakes Awake*, MIT Press, 1982.
10 Corliss, W.R., *Handbook of Unusual Natural Phenomena*, Sourcebook Project 1977.
11 Corliss, W.R., *The Unexplained*, Bantam, 1976.
12 *The Times*, 8 January 1987.

CHAPTER 2 THE CASCADES CATALYST

1 Arnold, Kenneth and Palmer, Ray, *The Coming of the Saucers*, 1952.
2 Arnold, Kenneth, 'How It All Began', *Proceedings of the First International UFO Congress* (1977), Warner Books, 1980.
3 Steiger, Brad (ed.), *Project Blue Book* (1976), Ballantine 1987 edition.
4 Arnold, Kenneth, 1977, *op. cit.*
5 *Ibid.*
6 Beer, Lionel, in *Phenomenon* (eds H. Evans and John Spencer), Futura, 1988.
7 Cullen, Regina, 'Restoring Arnold's Good Name', *UFO Brigantia*, March 1988.
8 Keyhoe, Donald E., *Flying Saucers from Outer Space*, Tandem, 1970.
9 Evans, Hilary, in *UFOS: 1947–1987* (eds H. Evans and J. Spencer), Fortean Times, 1987.
10 Arnold and Palmer, 1952, *op. cit.*
11 Watson, Nigel, in *UFOS: 1947–1987*, *op. cit.*

12 Quoted by Jacques Vallee in *Passport to Magonia* (1970), Tandem edition 1975.
13 From Herman of Laon's *De Miraculis Sanctae Mariae Laudunensis*, translated from the Latin by Jeremy Harte, published privately by him as *The Dragon of Christchurch* in 1985. Harte appends some exceptionally valuable notes on dragons in medieval literature, and makes special reference to them as possible geophysical phenomena, with particular reference to the earth lights theory. This excellent work has been used for most of the dragon references in this outline history of UFOs.
14 Bird, Geoff, 'Mining Lights', *Ley Hunter*, 102, 1987.
15 Mac Manus, Dermot, *The Middle Kingdom* (1959), Colin Smythe edition 1979.
16 Liljegren, Anders, and Svahn, Clas, in *UFOS: 1947–1987*, *op. cit.*
17 Jung, C.G., *Flying Saucers – A Modern Myth of Things Seen in the Sky* (1958), RKP edition 1959.
18 Quoted in Harte, 1985, *op. cit.*
19 Lagarde, F., in *Flying Saucer Review*, vol.14, no.4, 1968.
20 Michell, John, *The View Over Atlantis*, Sago Press, 1969.
21 Keel, John, *UFOS – Operation Trojan Horse* (1970), Abacus edition 1973.
22 Devereux, Paul, and York, Andrew, 'Portrait of a Fault Area', *The News* (now *Fortean Times*), 11 and 12, 1975.
23 Devereux, Paul, and York, Andrew, *Ley Hunter*, 66, 67, 68, 1975.
24 Persinger, M.A., and Lafrenière, Gyslaine, *Space-Time Transients and Unusual Events*, Nelson Hall, 1977.
25 Finkelstein, D., and Powell, J., 'Earthquake Lightning', *Nature*, 228, 1970.
26 Forshufvud, R., 'UFOs – A Physical Phenomenon', *Pursuit*, 50, 1980.
27 Devereux, Paul, *Earth Lights*, Turnstone Press, 1982.
28 Tributsch, Helmut, *When the Snakes Awake*, MIT Press, 1982.
29 Randles, Jenny, *The Pennine UFO Mystery*, Granada, 1983.
30 Devereux, Paul, McCartney, Paul, Robins, Don, 'Bringing UFOs Down to Earth', *New Scientist*, 1 September 1983.
31 Persinger, Michael, in interview with Dennis Stacy, *Fortean Times*, 42, 1984.
32 Associated Press despatch in *Trenton Times*, 31 March 1980.

CHAPTER 3 BRITISH LIGHT PHENOMENA

1 Magin, Ulrich, 'Highland Mysteries', *INFO*, 51, 1987.
2 MacGregor, Alasdair Alpin, *The Ghost Book*, Hale, 1955.
3 *Ibid.*

4 *Daily Mail*, 1 January 1913.
5 *Impartial Reporter and Farmers' Journal*, January 1913.
6 Personal communication from George Sandwith, July 1983.
7 *Mid-Ulster Mail*, 14 December 1912.
8 *Manchester Guardian*, 9 February 1905.
9 Quoted in *Daily Mirror*, 14 February 1905.
10 *Daily Mail*, 13 February 1905.
11 *Daily Mirror*, 13 February 1905.
12 *Daily Mirror*, 16 February 1905.
13 *Daily Mirror*, 14 February 1905.
14 *Daily Mirror*, 16 February 1905.
15 McClure, Kevin and Sue, *Stars and Rumours of Stars*, private, 1980.
16 *Manchester Guardian*, 14 February 1905.
17 Bord, Janet and Colin, *Flying Saucer Review*, March 1979.
18 Fry, Margaret, reports in *Northern UFO News* (ed. Jenny Randles), 116, 117, 118, 120, 121, 122, 123, 128, 1985–7.
19 'Llowarch', *Weird Wonders of Wales*, Cambrian News publications.
20 Devereux, Paul, *Places of Power*, Blandford Press (in preparation).
21 Mullard, Jonathan, 'The Linley Lights', *Ley Hunter*, 97, 1985.
22 Clarke, David, and Oldroyd, Granville, *Spooklights – A British Survey*, private, 1985.
23 Clarke, David, 'Cotswolds Lights', *Ley Hunter*, 99, 1985.
24 *Royal Leamington Spa Courier and Warwickshire Standard*, 16 February 1923.
25 *Birmingham Post*, 17 February 1923.
26 *Birmingham Gazette*, 20 February 1923.
27 *Leamington, Warwick, Kenilworth & District Morning News*, 28 January 1924.
28 *English Mechanic & World of Science*, 3 October 1919.
29 *Notes & Queries*, 4 April 1891.
29 PRO Admiralty, 6 December 1915.
31 PRO Admiralty, 31 March 1916.
32 Baring-Gould, S., *A Book of Dartmoor*, Methuen, 1900.
33 Personal comment from Chris Castle, 1982; personal communication, 1988.

CHAPTER 4 PROJECT PENNINE

1 Bellamy, Rex, *The Peak District Companion*, 1981.
2 Still, Ernie, report for BUFORA.

3 Randles, Jenny, *Pennine UFO Mystery*, Granada, 1983
4 Investigated by Nigel Watson of UFOIN.
5 Reported in the *Batley News*.
6 Randles, Jenny, *op. cit.*
7 Bennett, Paul, 'Calverley Connections', *Ley Hunter*, 103, 1987.
8 Tributsch, H., *When the Snakes Awake*, MIT Press, 1982.
9 McClennan, Alex, *The Lost World of Agharti*, 1983.
10 'Ghost in Gaping Gill', *Craven Pothole Club Journal*, 1964.
11 Roberts, A. (ed.) *UFO Brigantia*, files.
12 *Flying Saucer Review*, July 1957.
13 *Flying Saucer Review*, November 1956.
14 Harris, J., *The Ghost Hunter's Road Book*, Muller, 1968.
15 MacGregor, A.A., *The Ghost Book*, Hale, 1955.
16 Mitchell, W. R., *Ghosts of Yorkshire*, Dalesman, 1987.

CHAPTER 5 AMERICAN SPOOKLIGHTS

1 Tiede, Tom, 'Marfa Lights Lore: Texas Fireflies?' *Saginaw News*, 13 October 1985.
2 Miles, Elton, *Tales of the Big Bend*, Texas A&M University Press, 1976.
3 Stacy, Dennis, 'Ghost Lights Remain as Marfa's Top Attraction', *Fort Worth Star-Telegram*, 5 August 1984.
4 Simon, Ted J., 'Marfa's Ghost Lights Persist', *Lubbock Avalanche Journal*, 27 October 1985.
5 Kenney, Pat, and Wright, Elwood, '*The Enigma of Marfa – An Unexplained Phenomenon*', unpublished account, with attached note by Pat Kenney dated 15 March 1979, attesting to the truth of the account, and stating that the close sighting was recorded independently by both himself and Wright within one hour of the sighting.
6 Hanners, David, 'Marfa Lights Convince Pair of Geologists', *Dallas Morning News*, 4 July 1982.
7 Stacy, Dennis, 1984, *op. cit.*
8 Stewart, Richard, 'Bragg Road is a Spooky Place to Go', *Texas Chronicle* 31 October 1986.
9 Kubrin, David, personal comment 1986; subsequent personal communication.
10 Kubrin, Karen, personal communication, February 1988.
11 Kubrin, David, Devereux, Paul, McCartney, Paul, 'Pinnacles Light', *Ley Hunter*, 101, 1986.
12 Corliss, W. R., *The Unexplained*, Bantam, 1976.
13 Cantor, Patricia, 'The Brown Mountain Lights', *Ley Hunter*, 100, 1986.
14 I made this point about an inner, seething nature to light balls in 1982 in *Earth Lights*.

15 Mansfield, George Rogers, *U.S. Geological Survey Circular 646*, 1971 (reproducing press notice 14328, 1922).
16 Cantor, Patricia, 1986, *op. cit.*
17 Anon, *Literary Digest*, 87, 1925.
18 Harden, John, *Tar Heel Ghosts*, University of North Carolina Press, 1954.
19 Gibson, E., *Strange Hand of Fate*, USA 1967.
20 Edwards, Frank, *Strange World*, Bantam, 1964.
21 Quoted by Curtis Fuller in *Fate*, 15, 1962.
22 Kaczmarek, Dale, 'The Ozark Spooklight', *INFO*, 45, 1984.
23 Edwards, Frank, 1964, *op. cit.*
24 Rutledge, Harley D., *Project Identification*, Prentice-Hall, 1981.
25 Rutledge, Harley D., interviewed in *The Null Report*, WBAI 99, 5 FM, 20 April 1987.
26 Salisbury, Frank, B., *The Utah UFO Display*, Devin-Adair, 1974.
27 Long, Greg, 'Yakima Indian Reservation Sightings', *MUFON UFO Journal*, 166, 1981.
28 Devereux, Paul, 'New Jersey Light', *Ley Hunter*, 98, 1985.
29 *Science Digest*, July 1982.
30 Coleman, Loren, *Curious Encounters*, Faber & Faber, 1985.
31 Atwater, P.M.H., 'Pine Bush Lights', *Ley Hunter*, 103, 1987.
32 Hynek, J. Allen, Imbrogno, Philip J., and Pratt, Bob, *Night Siege – The Hudson Valley Sightings*, Ballantine, 1987.

CHAPTER 6 LIGHTS WORLDWIDE

1 Strand, Erling, *Project Hessdalen: Final Technical Report. 1984. Part 1*, Norway, 1985.
2 Hansen, Kim 'UFO Casebook', in *UFOS: 1947–1987* (eds H. Evans and J. Spencer), Fortean Times, 1987.
3 Røed, Odd-Gunnar, and Strand, Erling, 'Project Hessdalen Update', paper given to the BUFORA *Congress 87*, London, 1987.
4 Evans, Hilary, 'Northern Lights', *Magonia*, 14, 1983.
5 Havik, Leif, 'More about the Hessdalen Phenomenon', *Bolide*, 3, 1987.
6 Evans, Hilary, 1983, *op cit.*
7 Mattsson, Dan, 'UFOs in Time and Space', *AFU Newsletter*, 27, 1985.
8 Mattsson, Dan, and Mattsson, Carl-Anton, 'Kolmarden – A Swedish UFO Window', *UFO-Sverige Aktuellt*, vol.3, no.4.
9 Kingsley, Mary, *Travels in West Africa*, Virago 1982 edition.
10 Bauer, N.W., 'A Mystery Unsolved – the Story of the Min Min Light', *Bulletin of the Royal Geographic Society of Australia*, January 1982.
11 *Ibid.*
12 'Nullarbor UFO', *Nexus*, 4, 1988.

13 Kingsley, William, 'Ghost Lights', *Pursuit*, 4, 1986.
14 Morrison, Tony, *The Mystery of the Nasca Lines*, Nonesuch Expeditions, 1987.
15 Luce, Cynthia Newby, 'Brazilian Spooklights', *Fortean Times*, 49, 1987.
16 Luce, Cynthia Newby, personal communication to Andy Roberts, February 1988.
17 Mac Manus, Dermot, *The Middle Kingdom* (1959), Colin Smythe edition, 1973.
18 Corliss, W.R., *Handbook of Unusual Natural Phenomena*, Sourcebook Project, 1977.
19 Brunton, Paul, *A Search in Secret Egypt*, 1936.
20 Robson, G., *Marine Observer*, 25, 1955.
21 Huntingdon, Ellsworth, *Monthly Weather Review*, 28, 1900.
22 Rhys, John, *Celtic Folklore* (1901), Wildwood House edition, 1980.
23 Bord, Janet and Colin, *The Secret Country*, Elek, 1976.
24 Maxwell, George, *In Malay Forests* (relevant section reproduced in *Bolide* 2, 1987).
25 Sherrard, Philip, *Athos – The Holy Mountain*, Overlook Press, 1982.
26 Collins, Andrew, *The Knights of Danbury*, Earthquest Books, 1985.
27 Bayley, Harold, *Archaic England*, Chapman & Hall, 1919.
28 Collins, Andrew, 'Mount Athos', *Ley Hunter*, 104, 1987.
29 Sherrard, Philip, 1982, *op. cit.*
30 Deland, Inga, personal communication, 1987.
31 Clarke, David, 'Haunted Hills and Spooky Spots', *UFO Brigantia*, 22, 1986.
32 Devereux, Paul, and Thomson, Ian, *The Ley Hunter's Companion*, Thames & Hudson, 1979.
33 Devereux, Paul, *Earth Lights*, Turnstone Press, 1982.

CHAPTER 7 EARTH BORN

1 Lilwall, R.C., *Seismicity and Seismic Hazard in Britain*, HMSO, 1976.
2 Méreaux, Pierre, *Carnac – une porte vers l'inconnu*, Laffont, 1981.
3 Devereux, Paul, *Places of Power*, Blandford Press (in preparation).
4 Clarke, David, 'Cotswold Lights', *Ley Hunter*, 99, 1985.
5 *Leamington Chronicle*, 8 February 1924.
6 Clarke, David, and Oldroyd, Granville, *Spooklights – A British Survey*, private, 1985.
7 Tributsch, H., *When the Snakes Awake*, MIT Press, 1982.
8 'Earth Emits Ghostly Light', *Science Digest*, July 1982.
9 Hanners, David, *Dallas Morning News*, 4 July 1982.
10 Gibson, E., *Strange Hand of Fate*, *op. cit.* 1967.
11 Persinger, Michael A., personal communication, February 1988.

12 Except for *Ley Hunter* 101, 1986.
13 Persinger, M.A., and Derr, J. S., 'Relations between UFO Reports within the Uinta Basin and Local Seismicity', *Perceptual and Motor Skills*, 60, 1985.
14 Derr, J.S., and Persinger, M.A., 'Luminous Phenomena and Earthquakes in Southern Washington', *Experientia*, 42, 1986.
15 Evans, Hilary, 'Northern Lights', *Magonia*, 14, 1983.
16 Strand, Erling, *Project Hessdalen: 1984 – Final Technical Report, Part 1*, Norway, 1985.
17 Strand, Erling, 'Thoughts on the Hessdalen Phenomenon', *Bolide*, 1, 1986.
18 Randles, Jenny, *The UFO World '86*, BUFORA, 1986.
19 Mattsson, Dan, 'UFOS in Time and Space', *AFU Newsletter*, 27, 1984.
20 Mattsson, Dan, and Persinger, M.A., 'Positive Correlations Between Numbers of UFO Reports and Earthquake Activity in Sweden', *Perceptual and Motor Skills*, 63, 1986.
21 'Nullarbor UFO', *Nexus*, 4, 1988.
22 Peterson, I., 'The Light Side of Rock Fractures', *Science News*, 14 June 1986.
23 Persinger, M.A., personal communication, June 1988.
24 Brady, Brian T., and Rowell, Glen A., 'The Laboratory Investigation of the Electrodynamics of Rock Fracture', *Nature*, 29 May 1986.
25 Derr, John S., 'Luminous Phenomena and their Relationship to Rock Fracture', *Nature*, 29 May 1986.

CHAPTER 8 EARTH LIGHTS AND CONSCIOUSNESS

1 Constable, Trevor, *The Cosmic Pulse of Life*, Spearman, 1976.
2 Howarth, R., Mantle, P., and Roberts, A., 'A Soapy Abduction', *UFO Brigantia*, 32, 1988.
3 Keul, Alex, and Phillips, Ken, 'Assessing the Witness: Psychology and the UFO Reporter', in *UFOs: 1947–1987* (eds H. Evans and J. Spencer), Fortean Times, 1987.
4 Keul, Alex, and Phillips, Ken, 'The UFO – An Unidentified Form of Creativity?', paper presented to BUFORA's *Congress 87*, London, 1987.
5 Quoted by Kim Hansen in *UFOs: 1947–1987*, *op. cit.*
6 Persinger, M.A., 'Clinical Consequences of Close Proximity to UFO-Related Luminosities', *Perceptual and Motor Skills*, 56, 1983.
7 Devereux, Paul, *Places of Power*, Blandford Press (in preparation).
8 Persinger, M.A., 'Religious and Mystical Experiences as Artifacts of Temporal Lobe Function: A General Hypothesis', *Perceptual and Motor Skills*, 57, 1983.

9 Hopkins, Bud, *Intruders*, Random House, 1987.
10 Strieber, Whitley, *Communion*, Century, 1987.
11 Wootten, Mike, 'Communion Two', *BUFORA Bulletin*, 28, 1988.
12 Referred to in many of her writings; for example *Science and the UFOs*, Blackwell, 1985.
13 Jung, C.G. *Flying Saucers – A Modern Myth of Things Seen in the Sky*, (1958), RKP edition 1959.
14 Persinger, M.A., 'The Tectonogenic Strain Continuum of Unusual Events', *Perceptual and Motor Skills*, 60, 1985.
15 Persinger, M.A., and Cameron, Robert A., 'Are Earth Faults at Fault in some Poltergeist-Like Episodes?', *Journal of the American Society for Psychical Research*, 80, 1986.
16 Grant, James Shaw, *The Gaelic Vikings*, James Thin, 1984.
17 Gearheart, L., and Persinger, M.A., 'Onsets of Historical and Contemporary Poltergeist Episodes Occurred with Sudden Increases in Geomagnetic Activity', *Perceptual and Motor Skills*, 62, 1986.
18 Persinger, M.A., 'Intense Paranormal Experiences Occur During Days of Quiet Global Geomagnetic Activity', *Perceptual and Motor Skills*, 61, 1985.
19 Schaut, G.B. and Persinger, M.A., 'Global Magnetic Activity During Spontaneous Paranormal Experiences: A Replication', *Perceptual and Motor Skills*, 61, 1985.
20 Persinger, M.A., 'Spontaneous Telepathic Experiences from *Phantasms of the Living* and Low Global Geomagnetic Activity', *Journal of the American Society for Psychical Research*, 81, 1987.
21 Further discussion of these matters occurs in *Earthmind* by Paul Devereux (with John Steele and David Kubrin) Harper & Row (USA) – in preparation.
22 Findler, Gerald, *Lakeland Ghosts*, Dalesman, 1970.
23 Harris, John, *The Ghost Hunter's Road Book*, Muller, 1968.
24 *The Unknown*, July 1987.
24 *The Mirror of Literature, Amusement and Instruction*, 11 April 1835.
26 Rickman, Phil, *Mysterious Lancashire*, Dalesman, 1977.
27 Whitaker, T. W., *Lancashire's Ghosts and Legends*, Hale, 1980.
28 Cheney, Margaret, *Tesla: Man Out of Time*, Laurel, 1981.
29 Derr, John S., personal communication, August 1986.
30 Lofthouse, Jessica, *North Country Folklore*, Hale, 1976.
31 Salisbury, Frank B., *The Utah UFO Display*, Devin-Adair, 1974.
32 Havik, Leif, 'More About the Hessdalen Phenomenon', *Bolide*, 3, 1987.
33 Rutledge, Harley D., *Project Identification*, Prentice Hall, 1981.

INDEX

Page numbers in *italics* refer to illustrations

INDEX

INDEX